MORE MAKERS OF AMERICAN MACHINIST'S TOOLS

PART TWO
OF A HISTORICAL DIRECTORY OF
MAKERS AND THEIR TOOLS

Kenneth L. Cope

ASTRAGAL PRESS
Mendham, New Jersey

Copyright © 1998 by The Astragal Press

All rights reserved. No part of this book may be reproduced or transmitted in any form or by any means, electrical or mechanical, including photocopying, recording, or by any information or retrieval system, without the written permission of the Publisher, except where permitted by law.

Library of Congress Catalog Card Number 98-71297
International Standard Book Number 1-879335-

Published by
THE ASTRAGAL PRESS
5 Cold Hill Road, Suite 12
P.O. Box 239
Mendham, New Jersey 07945-0239

Manufactured in the United States of America

DEDICATION

This book is dedicated to all collectors of machinist's tools, wherever located. Good hunting, and if you find some great tool not listed in this book or its predecessor, please drop me a line.

TABLE OF CONTENTS

Acknowledgements *i*

Introduction *ii*

The Makers of American Machinist's Tools:
An Alphabetic Listing, with illustrations of their products *1 - 111*

Appendix I:
The Development of Fixed Caliper Gages, 1862-1878 *112 - 114*

Appendix II: Darling & Schwartz *115-117*

Appendix III: John Coffin Obituary *118-119*

Appendix IV: Sensitiveness of Touch *120 - 121*

Appendix V: Micrometer Calipers of 1917 *122 - 129*

Appendix VI: Memories of an All-Round Machinist *130 - 133*

Appendix VII: 1915 Union Caliper Co. Catalog *134 - 149*

Additional Machinist's Tool Patents *150 - 165*

Patent Drawings *166 - 185*

ACKNOWLEDGEMENTS

Only the unselfish help of other collectors have made this book, and its predecessor, possible. I wonder if any other group would be so willing to furnish information, send rare and delicate publications and tools through the mail, and offer their expertise in any way deemed helpful. I doubt it. Special thanks are tendered to:

Ken Kranzusch, Montgomery, AL

Robert Lang, Minneapolis, MN

Roy Schaffer, Columbia, MD

Roger Smith, Athol, MA

This book contains important information from the files of William R. Robertson. In addition to my thanks to him for so much help, he wishes to thank the following who contributed information to him:

Edwin A. Battison

Burton Cohen

Hunter Pilkington

Karl Sanger

Roy Schaffer

Roger Smith

Philip E. Stanley

American Precision Museum

Early American Industries Association

Linda Hall Library, Kansas City, MO

Smithsonian Institution

Ken Cope
Milwaukee, WI

INTRODUCTION

When *Makers of American Machinist's Tools* was published in 1994, it was with the sure knowledge that it could not possibly be complete. As a pioneering work, there was no previous data to build on, nor any means of knowing what might be missing.

As hoped, the publication of the first book stimulated a good deal of interest in the subject of machinist's tools and encouraged many collectors to come forth with tool makers and tools which had been unknown, or inadequately known, in 1994. Many of them supplied catalog copies, tools, and instruction sheets which form the basis for much of this book. At the same time, research continued, centering on publications such as *American Machinist, Iron Age,* and *Machinery*. A number of previously unrecognized makers were located in those pages.

This is not to imply that this book has every maker of American machinist's tools not found in the first. The 200 entries in this book, added to the 300 or so in the first, cover a large percentage of the total. However, any list of makers can never be all-inclusive. If we take the term "maker" literally, nearly every machinist who lived is a maker of a few machinist's tools. We can only concentrate on commercial makers and hope to cover those of interest to today's collectors.

Here in Milwaukee, we see a good example of confusion caused by tools made by individuals. Calipers, surface gages, rule stands and vises, all of the same pattern, but all marked with different maker's names abound at local flea markets. These tools were made at the local vocational school which, for many years, worked directly with industry in their apprentice programs. Making basic machinist's tools to drawings furnished by the school was a part of most machinist apprentice programs. Many of these tools are so well made that they are often assumed to be of commercial origin.

A few of the entries in this book are included to correct errors found in the first volume of *Makers of American Machinist's Tools*. Others are expansions of entries in the first book, especially when an illustration of the tool(s) became available.

The combination of the three books, *American Machinist's Tools: An Illustrated Directory of Patents; Makers of American Machinist's Tools;* and *Makers of American Machinist's Tools, Volume II* furnishes collectors and students of American tools with solid reference material on a long neglected subject. I hope it will prove useful in preserving an important part of our industrial heritage.

Kenneth L. Cope
Milwaukee, Wisconsin

Alphabetic Listing of the Makers of American Machinist's Tools

with illustrations of their products

ALLIANCE TOOL CO., Alliance, OH

Maker, beginning in 1929, of a line of toolmaker's V-blocks, parallels, and planer jacks.

AMERICAN EVER READY CO., New York, NY

Maker, in 1912, of a revolution counter "shaped somewhat like a revolver and about the same size." Pulling the trigger started the count and releasing it stopped the count. At least one example is known.

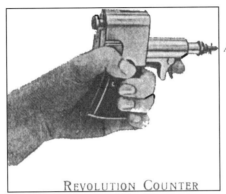

American Machinist 1912

AMERICAN STEAM GAGE & VALVE MFG. CO., Boston, MA

Maker, in 1908, of the direct reading speed indicator shown below.

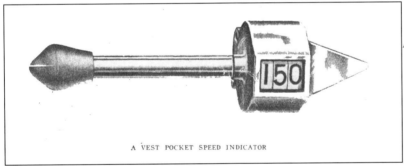

American Machinist 1908

AMES, NATHAN, Saugus, MA

Inventor of the Ames Universal Square patented July 6, 1852. An 1855 article in "Scientific American" magazine described the tool and noted that "never before this brought before the public", strongly suggesting that Ames had never manufactured the square. The first known maker, GAVIN HOLLIDAY of Lynn, MA, began production about 1856. By 1859, the patent rights were owned by J.R. BROWN & SHARPE who offered it in that year's catalog.

(See next page for illustration.)

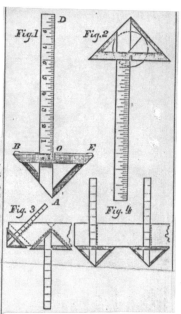

Scientific American 1855

ANROS CO., P.S., North Tonawanda, NY

Maker, in 1910, of a machinist's and draftsman's beam compass. The tool was "complete with pen, pencil, needle points and 12-in. beam, postpaid for $1.00."

ART TOOL CO., Bridgeport, CT

Maker, in 1921, of base blocks which converted vernier calipers into height gages, and toolmaker's combination square and V-blocks accurate to 0.0001".

ASHCROFT MFG. CO., New York, NY

Founded about 1860 to make steam gages, the company was bought by Manning, Maxwell & Moore, a large machinery sales house, in 1873. The firm continues to operate as a subsidiary of Dresser Industries. Maker, in 1912, of a sheet metal thickness gage, patented by Frederick Blanchard and E.B. Crocker on May 28, 1912, and sold through Manning, Maxwell & Moore at least as late as 1921.

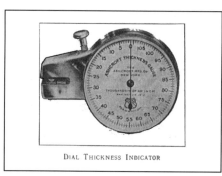

American Machinist 1912

DIAL THICKNESS INDICATOR

ATLAS BALL CO., Philadelphia, PA

Maker of ATLAS ball gages in 1913. The gages, for measuring internal diameters, were furnished in sets and consisted of an accurate ball, guaranteed to within 1/10,000 of an inch, welded to a knurled handle.

ATLAS TOOL CO., New York, NY

Maker of surface gages with a unique fine adjustment mechanism as shown below. The gages are well designed but rather crudely made. Nearly identical gages are known marked SANDOW TOOL CO., USA; others are completely unmarked.

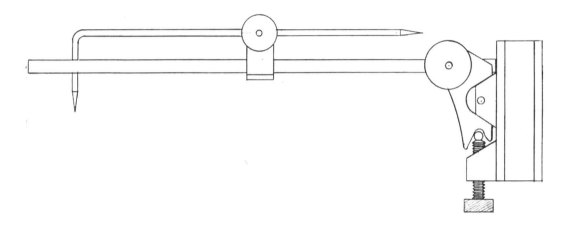

AVERY & CO., WILLIAM, Foxboro, MA

Maker, in 1910, of an inside thread-tool setter and center gage. It is described as: "a tool which enables one to reset the inside threading tool after it has been reground. One end has a male V which engages the partly finished internal thread; exactly one inch from this a female V is turned in the rod. With the male V meshing with the internal thread the tool is set to the female V, with the assurance that it will catch the thread correctly."

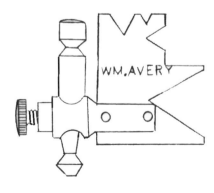

BARKER, ALFRED, Richfield Springs, NY

Inventor and maker of BARKER'S beam caliper, patented May 2, 1893. The patent covered a clever design that allowed the outer jaw to slide between two locating pins on the beam. This feature gave the machinist a means to change inside to outside measurements (or vice versa), a common need when making a shaft to fit a bore. As the ad below shows, the tools were sold through J.F. Getman as sole agent.

Scientific American 1893

BARSANTEE, JAMES W., East Somerville, MA

Inventor and maker of quick adjusting calipers, illustrated below. The tools are marked J.W. BARSANTEE, PAT. JUNE 9, 1874.

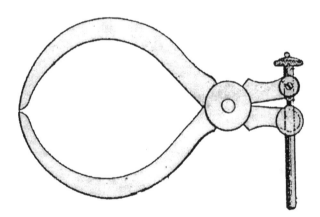

BATES, GEORGE A., Chicago, IL

Maker, about 1894, of an indicating square for which he was granted a patent on January 2, 1894. The only known example is marked with the patent date on the stock and D.B.& S. on the blade. The D.B.& S. mark means, of course, that the squares were either made by Bates from commercially purchased squares or were made by DARLING, BROWN & SHARPE under Bates' patent. The former appears most likely.

(See next page for illustration.)

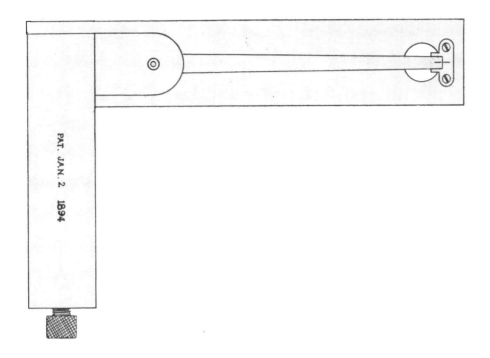

BEARD TOOL CO., L.O., Lancaster, PA

Operated by Lawrence O. Beard, the company was the maker, beginning in 1922, of a limited range Self Aligning and Centering Inside Micrometer Gauge and Heighth Gauge. The tool is an inside micrometer with no provision for extension rods and converts into a height gage when mounted in a small base attachment. The specimen examined is heavily marked with its measuring range, $3\frac{3}{4}"$ to $4\frac{1}{2}"$, which was the longest offered. Other sizes were $2\frac{1}{2}"$ to 3" and 3" to $3\frac{3}{4}"$. Tools examined are marked PAT APPLD FOR and therefore made before Beard received his patents on April 24, 1923, and July 17, 1923. A unique feature, claimed in the first patent, is the graduation of the sleeve from 0-50 in the clockwise direction and the spindle from 0-50 counter-clockwise. The reading could therefore be taken from either the sleeve or the spindle.

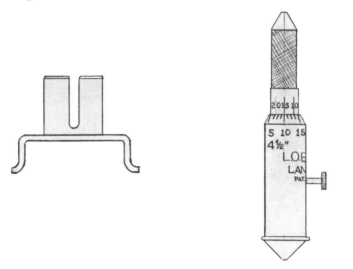

BELLAMY, CHARLES L., New York, NY

Buyer, in 1881, of the machinist's tool manufacturing business of JAMES D. FOOT. A January 1881 notice in "American Machinist" magazine stated that Bellamy was the originator of the tools made by Foot. Bellamy also patented a centering, bevel and try square November 14, 1882, which was offered by HUGH E. ASHCROFT in 1887 as the ACME center square. (See Ashcroft entry in *Makers of American Machinist's Tools.*)

BEMIS, S.C., Wilimansett, MA

Founded by Stephen C. Bemis about 1825. In 1828, he hired Amos Call as an apprentice and, in 1844, moved to Springfield, MA, where they formed the partnership of BEMIS & CALL. Some time before the partnership was formed, Bemis made a line of wing dividers and double calipers. One of the latter is shown below. The same tools with a BEMIS & CALL mark are also found.

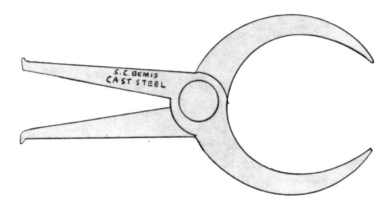

BEMIS & CALL CO., Springfield, MA

Formed in 1844 by Stephen C. Bemis and his former apprentice, Amos Call, as BEMIS & CALL. The firm was reorganized as the BEMIS & CALL CO. in 1855 and as the BEMIS & CALL HARDWARE & TOOL CO. in 1868. However, the company announced that they would continue to mark tools BEMIS & CALL CO.

An 1855 list of Springfield manufacturers stated: "manufactures dividers, compasses, calipers, punches, pocket squares, bevils (sic) and gauges, using up annually 10 tons of iron and two tons of steel, of the aggregate value of $4,000. Employs 23 hands, has been engaged in this business ten years and produces an annual amount of articles valued at $10,000."

The below 1871 advertisement from "Iron Age" magazine shows that they continued to make the same tools as in 1855, adding wrenches, hammers, saw-sets and steelyards. When the entry for *Makers of American Machinist's Tools* was written, it was believed that jointed calipers and small squares and bevels were the firm's only machinist's tool offerings.

Other types, shown below, are now known. The caliper square is very similar to the DARLING, BROWN & SHARPE calipers of the period; calibrated in 64ths on one side and 100ths on the other. The finely adjustable depth gage, offered in the 1899 F.W. Gesswein catalog, is a unique and extremely well made tool. The depth gage is marked PAT PENDING, but no patent has been found. (For more information, see BEMIS & CALL HARDWARE & TOOL CO. entry in *Makers of American Machinist's Tools* and A. CALL in this volume. *See illustration next page.*)

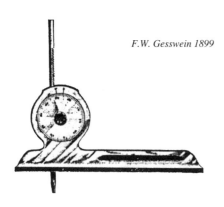

F.W. Gesswein 1899

FIG. 3085.

Price, $1 85 each.

The Depth Gauge consists of a frame with a base broad enough to support it in an upright position, the dial is rotated by a knurled nut at the back, concentric with it is a fine gear meshing with the teeth on the rod so that one complete rotation of the dial moves the rod one inch.

The dial is graduated in hundredths, the frame is provided with a vernier for reading in five-thousandths of an inch. The rod is of sufficient length for measurements of two inches.

We claim this tool to have a rapid and accurate adjustment and fine finish; which will commend it to tool makers, die sinkers, machinists, and fine workmen in general.

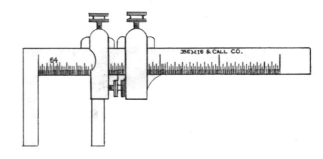

Iron Age 1871

BERG MFG. CO., E.G., Gardner, MA

Operated by Emil G. Berg, the company made inside micrometer sets similar to the #124 inside micrometers made by the L.S. STARRETT CO. Berg was granted a patent for a means to take up thread wear of micrometers on December 18, 1923, which may have been used on the inside micrometers. Date of operation is uncertain but probably about 1920-1930.

BERNHARDT, HERMAN V., Brooklyn, NY

Maker, circa 1891-1905, of a direct reading bench micrometer for which he received a patent September 15, 1891. The micrometer, shown below, was made with a 50 tpi screw and a dial divided into 200 divisions, giving .0001" per division. A second dial registered the number of turns of the main dial. When the readings of the two dials were added the user had a direct reading of the micrometer opening. Range was only .25 inches.

Bernhardt also made a special micrometer for jewelers and silversmiths to measure plate stock, wire, etc. The micrometer screw was made with 40 tpi and the dial was divided into 100 divisions, thus reading to 0.00025". He claimed that this "made very easy any translation of the wire gage size table into easy reading figures without facing a regiment of figures."

Bernhardt's patent of November 16, 1894, for a ratchet stop for micrometers, was bought and used by the BROWN & SHARPE MFG. Company *(See next page for illustration.)*

American Machinist 1904

A DIRECT READING MICROMETER.

BILLINGS & SPENCER CO., Hartford, CT

The following four pages from the BILLINGS & SPENCER CO. catalog of January, 1883, and eight pages from a ca.1911 catalog, add a good deal to our knowledge of this firm and its machinist's tools. Note the

PREFACE.

1883

THE BILLINGS & SPENCER COMPANY of Hartford, Conn., the Pioneers of Drop Forgings, as a regular and successful business, was organized in 1869, and in July, 1872, received from the Legislature of the State of Connecticut, a special Charter of Incorporation, conferring very favorable privileges, the Capital Stock being $150,000.00, with liberty to increase the same to $300,000.00.

In March, 1877, the Company reduced its Capital Stock to $125,000.00

Its Manufactory is located between Lawrence and Broad Streets, in the western part of the City of Hartford, and near the geographical center of the City, occupying a front of 312 feet on Lawrence Street, and 179 feet on Broad Street.

Its main factory is three stories high, covering an area of 40 by 130 feet, with Engine and Boiler Rooms. The Forge Shop is 82 by 100 feet. There are also several smaller buildings connected with the business of the Company on the premises.

As a specialty, the Drop Forging business is followed by very few concerns in the world, and there are none in the United States making in any comparison the amount and variety made by this Company.

The Manufactory is supplied with machinery and tools of improved make and description. In the Forge Shop are twenty-five drop hammers, ten presses, one atmospheric and four tilt hammers. The Machine Shop is furnished with lathes, planers, upright drill-machines, die-sinking machines, and a variety of Special Machines for use in manufacturing the several kinds of tools produced by the Company.

A full description of the almost endless variety of Drop Forgings, cannot be given in this limited space. Upwards of 2,000 different articles for parts of Guns, Pistols, Sewing Machines, Special Machinery, Machinists' Tools, Sewing Machine Shuttles, are among the Drop Forgings. A large variety of goods are also put upon the market in a finished state, by the Company, among which may be mentioned the following, viz.: Billings' Patent Screw Plates and Dies, Tap and Reamer Wrenches, Lathe Dogs in twelve sizes from 3/8 to 4 inches, Billings' Adjustable Pocket Wrench, Barwick Wrenches, Screw Drivers from best tool steel, Thread Cutting Tools, Billings' Improved double-action Ratchet Drills with sockets for using Morse Taper Shank Twist Drills, or the old-fashioned Square Shank Drills, five sizes Packer Ratchet Drills, about forty varieties of Sewing Machine Shuttles for the different makes of Sewing Machines, in this country and Europe. The Company employ 100 men, and its entire buildings are heated by steam and lighted by gas.

changes made in the surface gage by 1911 and the unusual .012" per inch die sinker's shrink rules. The combination caliper square and depth gage was covered by C.E. Billings' patent of August 11, 1903, and the pocket caliper by his patent of November 17, 1903. See listing in *Makers of American Machinist's Tools* for additional information.

C. E. Billings' Patent Surface Gauge.

1883

The hundred of uses of the surface gauge in modern mechanics make every improvement in its construction and adaptability of value to the practical, exact mechanic. One of the botherations of the ordinary surface gauge has been that to set it exact, dependence has been made wholly on the adjustability of a set screw, which demanded repeated trials on the "cut and try system." It is evident enough that it is possible to change this trial method to that where positive exactness shall be the rule, so that the carrying arm of the gauge points shall be as easily adjusted to exactness as the jaws of the spring calipers.

This Surface Gauge is drop-forged of bar steel, and finished in a thorough manner and hardened. It is much of the usual style, except the employment of two sliding snugs, connected by a screw encircled by an open spiral spring. The upper snug is split and is held in place at any position on the upright standard, by means of a simple thumb nut that clamps the split snug on the standard. This snug is connected to one below by means of a screw encircled by an open spiral wire spring. This lower snug carries the marking points, consisting of a piece of steel wire, which are held in the usual way by means of a thumb nut on a clamping screw.

In operation, any movement, up or down—along the line of the standard —or around its circumference, of the upper snug, will, of necessity, be accomplished by the lower snug in consequence of the connection of the screw; but the lower snug may be raised or be lowered by the connecting screw acting with the spiral spring, so that while the upper snug is held firmly in place by its binding screw, the lower snug, carrying the points, may be carefully and exactly adjusted to surface measurements, and when in position the tension of the spring and friction of the screw will hold the points exactly where they have been adjusted.

The advantages of this gauge are not confined to its close adjustment after the principal snug is fixed as closely as possible to the surface to be gauged, but comprehend also a swinging around of the gauge points to reach surfaces out of one direct line without disturbing the standard.

Cut represents Gauge, full size.

Price, . . $2.75

BILLINGS'
DROP-FORGED MACHINISTS' CLAMP.

1883

These clamps are forged from bar steel. They are a very useful tool for the machinist and tool maker, and one which, as a general thing, they have always been obliged to make for themselves. The old way of making these clamps was by using straight square bars, of the desired length and size, drilling, tapping, and putting in the screws. Necessarily there was a weak point, where the center screw passed through the pieces, and they were very apt to break at that point. The designer of this clamp, as shown in the cut, claims nothing particularly new, except it being drop-forged, and the stock distributed in a way to overcome the weak points named.

PRICE LIST.

No. 1, opening 1¼ inches, each, $1.50
" 2, " 2¼ " " 2.00
" 3, " 3¼ " " 2.50
" 4, " 4¼ " " 3.00

For Sale by

The Billings & Spencer Company,

HARTFORD, CONN., U.S.A.

Billings' New Hand Vise,

WITH

PARALLEL JAWS.

The cut represents this new Vise two-thirds size.

There has been great need for an improvement in Hand Vises, and particular attention has been given to adapting this Vise to the requirements of Linemen in the construction of Electric Light, Telegraph and Telephone Lines; also, for Machinists and Toolmakers.

They are drop-forged from solid Bar Steel, due regard having been given to distributing the metal in such a manner as to make it very strong where all the strength and resistance is required, and not have it too heavy or clumsy.

They are finished by Special Machinery, so that each part is interchangable and can be duplicated at nominal expense in case of necessity. They are hardened by a special process, which renders the jaws very tough and hard.

The jaws are parallel, which gives them a very strong grip, an opening being left in the Thumb-Screw head for extra purchase with lever, if wanted, and the action of the screw is such that the adjustment is very rapid.

PRICE LIST.

Vise, with loop, each,	$2.50
Vise, without loop, for machinists, each,	2.25
Vise Strap, best quality, each,	1.00

FOR SALE BY

The Billings & Spencer Co.
HARTFORD, CONN.

BILLINGS' IMPROVED POCKET CALIPER.

A very convenient pocket Caliper graduated on one side to 64ths, and on the other to 100ths. The tool opens 2 inches and is nicely finished. Furnished with or without Adjustment.

Price, with Adjusting Screw, ... $3 00
 "   without Adjusting Screw, ... 2 00

BILLINGS' COMBINATION CALIPER SQUARE, DEPTH GAUGE, AND 2-IN. AND 5-INCH STEEL RULES.

This combination tool is useful and convenient for machinists and toolmakers. Each rule is graduated in 64ths and 32ds, 16ths and 8ths.

The thumb nut is counterbored to receive a spiral spring which bears on the nut and washer, thus creating a moderate friction which, when the nut is loosened, facilitates the setting of the gauge to the desired measurement.

The tool is finished in the best possible manner, and opens 4 inches.

Price, each $4 00

BILLINGS & SPENCER CO.

BILLINGS' QUICK ADJUSTING DEPTH GAUGE.
For Die Sinkers and Tool Makers.

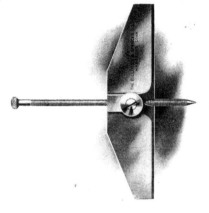

Length of Screw—3¾ in.

By making one-half turn of the nut and pressing on same, the screw threaded rod is released from the half nut and can then be moved freely up and down, releasing the pressure; the half nut is locked with the screw by action of the spring inside the nut. Fine adjustment can then be made by turning the screw. When adjustment is obtained screw the nut tight; this prevents the screw from turning.

Put up ½ dozen in each box.

Price, each $1.50

STANDARD STEEL RULES, AND DIE SINKERS' SHRINK RULES.

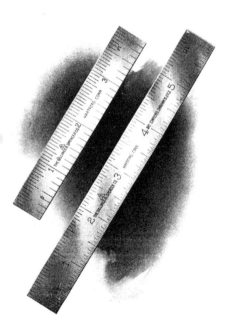

These Rules are carefully graduated and made to the standard widths.

The Shrink Rules are made specially for Die Sinkers, and are $\tfrac{3}{16}$ to the inch longer than the Standard Rules.

Price List.
Standard Rules.

2-inch Rule, 64ths and 32ds, 16ths and 8ths,	$0.25
4 " " " " " " " "	.45
5 " " " " " " " "	.55
6 " " " " " " " "	.65

Die Sinkers' Shrink Rules.

6-inch Rule, 64ths and 32ds, 16ths and 8ths,	$1.75
12 " " " " " " " "	3.00

BILLINGS & SPENCER CO.

BILLINGS' IMPROVED SCRATCH GAUGE.
with and without Screw Adjustment.

Without Screw Adjustment.   With Screw Adjustment.

The sliding head of this tool is drop forged, finished, hardened and polished. The scratch point is made of best tool steel and finely tempered. The entire tool is carefully fitted and highly finished.

Price List.

No. 1. 7 x 9/32, with screw adjustment, each	. .	$2 00
No. 2. 7 x 9/32, without " "	. .	1 50

BILLINGS' IMPROVED PATENT SURFACE GAUGE.

Made of steel throughout and finished in a thorough manner.

The base is drop forged, finished all over, hardened and ground to a true bearing. The pointer is made of the best grade of tool steel. The rest of the parts are made of machinery steel and thoroughly fitted and finished.

Price.

8-in. Standard . .	$2 75
9-in. " . .	2 75
12-in. " . .	3 00

14

JEWELERS' ANVIL, WITH BASE.

Square Anvil. Base, with Horn Anvil. Cutter or Hardie.

Cuts are ½ size.

These tools are especially designed for jeweler's or amateur's use. The base is drop forged from steel and is furnished either nickel plated or case hardened. The anvils and hardie are drop forged from suitable steel and properly hardened.

Price List.

Base, each	$0 70
Horn anvil, each	80
Square anvil, each	40
Hardie, each	40
Base, with the two Anvils and Hardie	2 00

BILLINGS' MICROMETER HOLDER.

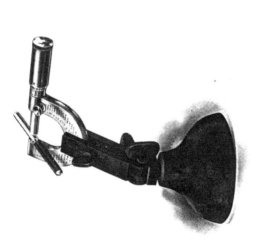

Cut is ½ size.

Weight is 1½ lbs.

It is made to tilt at different angles. The piece to be measured may be held in one hand, leaving the other free to adjust the Micrometer. A great convenience to every machinist.

Drop-forged of steel throughout.

Price List.

Holder, case hardened with Nickel Base	$2 50
" " " case hardened base	2 25

BLAZE, J. R., Whitestone, L.I., NY

Maker, in 1906, of "a V-block for spacing holes to be drilled in cylindrical and other shapes." The two blocks were connected by a rod with a series of grooves, spaced ¼" apart. One block could be locked into any of the grooves, while the other was moved into the desired position with a knob graduated to ¹⁄₁₂₈". A drill bushing was contained in the moveable block. The tool appears useful, but also very complex and expensive for its purpose.

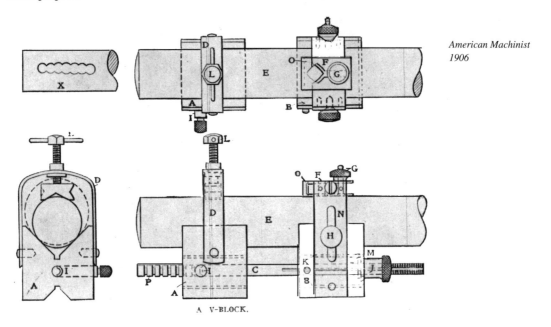

American Machinist 1906

A V-BLOCK.

BOETTCHER, F.W. GUSTAV, Milwaukee, WI

Inventor of a combination level, square and triangle patented September 2, 1884. The tool was offered in the 1889 FRASSE & CO. catalog, but the maker is not specified.

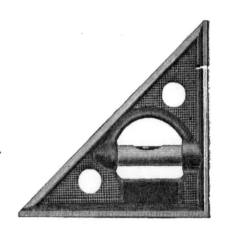

Patent Right-Angle Triangle Level.

4 inches. ....each, 75c.

Frasse & Co. 1889

BOWKER, M., Fitchburg, MA

Inventor and maker of Bowker's Conical Turning Gauge, patented August 1, 1865. The only known specimen, shown below, is in the Smithsonian collection.

Photo courtesy of the Smithsonian Institution, file number 308915.

BRATSCHI MFG. CO., Cleveland, OH

Maker of the Fuchs patent test indicator. It is not known if they made the tools before or after those made by the LEON FUCHS MFG. CO. of Dayton, OH. The known specimen is marked:

BRATSCHI MFG. CO.
FUCHS
PAT. NOV. 16, 1913
CLEVELAND, O. U.S.A.

BROWN, J.R., Providence, RI

Brown & Sharpe, for many years, published a very small picture of the original J.R. Brown vernier caliper, introduced in 1851, in the introduction to their catalogs. We could see the general appearance, but details were very hard to discern. A full size, original photo has become available and is shown below. As can be seen, the principle is unchanged from the later, more common vernier calipers, but many details are different. The very short jaws and the rather awkward thumb screw were soon improved. Mounting the vernier scale in line with the screw adjustment was actually a slightly more accurate method than that used on later production, but the lack of means for wear compensation more than offset any advantage of the in-line mounting.

Photo courtesy of the Smithsonian Institution, file number 321325.

BROWN & SHARPE, J.R., Providence, RI

The following June 1859 J.R. BROWN & SHARPE catalog is the earliest known. Only four pages in length, it shows the complete line of machinist's tools then offered. Note the early form of vernier caliper with the thumb release. Note also that it was offered in both German silver and steel. No surviving German silver calipers have been reported. See entry in *Makers of American Machinist's Tools*, first volume, for the firm's history.

J. R. BROWN & SHARPE,
MANUFACTURERS OF
MACHINE-DIVIDED
U. S. STANDARD RULES,
AND
MACHINISTS' TOOLS,
115 SOUTH MAIN STREET, PROVIDENCE, R. I.

STEEL CALIPER RULES.

The above cut is a fac-simile of one side of these rules. The other side is divided 12ths, 24ths, 48ths, 8ths, 14ths, and 28ths, on the outside, and upon the slide to 32ds & 64ths of inches. When closed they are three inches long. The caliper can be drawn out to measure two and a half inches. The thickness of the rule is one-tenth of an inch. The advantages claimed are their superior accuracy and durability. PRICE $2,00.

VERNIER CALIPER.

The above cut is a fac-simile of one side of the Vernier Caliper, which reads to thousandths of inches. On the other side are 64ths of inches, to read without a vernier. This instrument is furnished with both inside and outside calipers and points, to transfer the distance with dividers. It measures seven inches. An explanation of the Vernier accompanies each instrument. These instruments are made of German silver and steel. Those made of steel have the points tempered. PRICE, in Morocco case, $6,00.

WILLIS' ODONTOGRAPH.

This is an instrument recently invented by Prof. R. WILLIS, of Cambridge University, England, for describing the correct form for the teeth of wheels, and the templets and cutters used in making them. All wheels of the same pitch, but of different sizes, having their teeth drawn with this instrument, will run together correctly. No. 1 is used for drawing the teeth of small wheels by diametral pitch when only a single arc is required. Price $2,00, with drawing and directions for use. No. 2 is for drawing the teeth of larger wheels by circular pitch, where it is necessary to have separate arcs for flanks and faces. Price $3,50 with drawings and directions for use. Price per pair, $5,00.

PAPER SCALES.

The advantages of these Scales are—they expand and contract nearly the same as drawing paper, do not soil the work, and distances can be set off from them without the use of dividers. They are 18 inches long. Price $1 per set of six.
Series A contains 6 Scales of ⅛, ¼, ⅜, 1, 1½, and 3 in. to the foot, for Architects.
" B " 6 " " 3-32, ⅛, 3-16, ¼ and ⅜ in. to the ft. "
" C " 6 " " 10, 20, 30, 40, 50 & 60 parts to the in. for Engineers.

JAMES A. BALCH, General Agent.

Price List of Standard Rules and Machinists' Tools.

The No. will be found before "U. S. St'd," on each Rule.

No.	Prices.	Description
2	$3 00	24 inch steel rule divided to 32ds, 48ths, 50ths and 64ths of an inch.
12	1 50	12 " " " " ditto.
13	75	6 " " " " ditto.
17	50	4 " " " " ditto.
91	3 00	24 inch steel rule divided to 8ths, 16ths, 32ds and 64ths of an inch.
92	2 25	18 " " " " ditto.
93	1 50	12 " " " " ditto.
94	1 13	9 " " " " ditto.
95	75	6 " " " " ditto.
96	50	4 " " " " ditto.
97	38	3 " " " " ditto.
80	3 00	24 inch steel rule divided to 8ths, 10ths, 12ths, 14ths, 16ths, 20ths, 24ths, 28ths, 32ds, 48ths, 50ths, 64ths and 100ths of an inch.
81	2 25	18 " " " " ditto.
82	1 50	12 " " " " ditto.
83	1 13	9 " " " " ditto.
84	75	6 " " " " ditto.
85	50	4 " " " " ditto.
86	38	3 " " " " ditto.
5	3 00	24 inch steel rule divided to 48ths, 50ths and 64ths of an inch, and also for diameter and circumference.
33	1 50	12 " " " ditto.
61	2 37	12 " steel Geer Rule, divided to 18ths, 20ths, 22ds, 24ths, 26ths, 28ths, 30ths and 32ds, whole length.
78	1 67	12 " " " " " 6ths, 7ths 8ths 9ths, 10ths, 11ths, 12th, 14ths, 16ths, 18ths, 20ths, 22nds, 24ths, 26ths, 28ths, 30ths, 32ds, 34ths, 36ths, and 38ths of an in.—1 inch in each division.
63	4 00	24 inch Triangular boxwood rule, divided to scales of ⅛, ¾, 3-16, 3-32, ⅜, ¾, ½, 1, 1½, and 3 in. to the ft. and 16ths of in.
64	2 00	12 " " " " ditto.
73	2 00	12 " " " " on one edge to 10ths, 20ths, 30ths, 40ths, 50ths and 60ths of an inch.
74	1 50	8 " " " " ditto.
34	4 00	Bevel Protractor, with sliding arm, and half circle divided to degrees—with 10 in. sliding arm, $4.50. A very useful article for machinists.
90	2 00	3 inch steel Caliper Rule, like cut.—divided like No. 86.

SHRINK RULES.

| 87 | 3 50 | 24½ inch Steel Rule, divided like No. 80. |
| 88 | 2 00 | 24½ " Boxwood Rule, ditto. |

AMES' PATENT UNIVERSAL SQUARE.

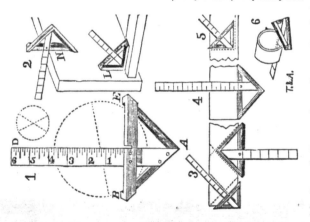

The tougne, D A, (Fig. 1,) being fastened, as it is, into the triangular frame, B A E, cannot be moved or knocked from its place,—in this respect, constituting a great improvement over the carpenters' Try-Square, T-Square and Miter, in common use. The instruments are both made of the best material, neatly finished, and perfectly true.

"As a **CENTER-SQUARE** alone, it is invaluable to every mechanic. *** In short, it combines, in a most convenient form, so many useful instruments, no mechanic's list of tools can well be complete without a Universal Square."—*Sc. Amer, Sep. 22, 1855.*

Fig. 1, explains its application as a CENTER-SQUARE. Put the instrument over the of sole, as the end of a bolt or shaft, with the arms, B A, E A, resting against the circumference, in which position one edge of the rule, A D, will cross the center. Mark a straight line in this position; apply the instrument again to another part of the circumference, and mark another line crossing the first. The point where the two lines cross each other, will be the center of the circle. Fig. 2, explains the application of the instrument, as a carpenter's TRY-SQUARE, M, and an OUTSIDE-SQUARE, L:—*Fig. 3*, as a MITER;—*Fig. 4*, as a T-SQUARE and a GRADUATED RULE;—*Figs. 5*, and *6*, as an OUTSIDE-SQUARE for drawing, and a T-SQUARE for machinists.

'This Square combines, in the most convenient form, *five different instruments;* viz: The TRY SQUARE, the MITER, the T-SQUARE, the GRADUATED RULE, and (what is entirely new) the CENTER SQUARE, for finding the center of a circle. The following are the retail prices:

No. 1, 6 in. blade, $1 75; 2, 8 in. $2 00; 3, 10 in. $2 25; 4, 12 in. $2 75.

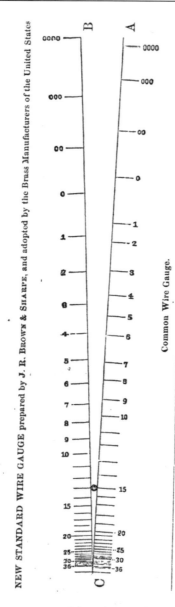

Improvement in the Wire Gauge.

To Manufacturers, Dealers and Consumers of Wire and Sheet Metals:

The undersigned, having for several years past been engaged in making Metal Rules and Gauges of various kinds, have, at the request of manufacturers and others, prepared a

STANDARD WIRE GAUGE,

with a new grade of sizes, which has met the approval of the principal Wire Drawers to whom it has been presented, and has been adopted by the Brass Makers as their Standard Gauge.

The want of uniformity in common Wire Gauges is well known, but if they all agreed with the published tables of sizes, there would still exist serious objections to their use, as the variation between different numbers is so irregular. This will be more clearly seen by reference to the diagram.

The two lines AC and BC meeting at C, represent the opening of an angular gauge. The divisions on the line AC, show the size of wire by the common gauge, those on the line BC by the new standard.

Wire to be measured by such a gauge, is passed into the angular opening till it touches on both sides, the division at the point of contact indicating the number. Thus, No. 15 old gauge, would be No. 13 by the new. The angular principle is used in the above cut, as it shows the difference between the old and new standard to the best advantage; it is proposed, however, to make gauges of different forms, but all to correspond with the sizes of the new standard.

The divisions on the line AC, it will be observed, are very irregular, while those on BC increase by a regular Geometrical Progression. This principle is thought by many who are conversant with the subject, to be the true one for the construction of a gauge, and when generally adopted by the manufacturers in this country, an effort will be made to introduce it in England.

The annexed tables show the actual dimensions of the old and new standards in decimal parts of an inch, U. S. Standard Measure, and also the difference between consecutive sizes of each gauge.

In order that the full benefit of the change may be realized, it is necessary that the gauges should be made to correspond exactly with the sizes given in the table, and feeling confident that we have accomplished this in a satisfactory manner, we submit the matter for your consideration. Very Respectfully,

J. R. BROWN & SHARPE,
115 South Main Street, Providence, R. I.

April, 1857.

No. of Wire Gauge	New Stand'd — Size of each Num. ber in decimal parts of an inch.	New Stand'd — Difference between consecutive Nos. in dec. parts of an in.	Old Stand'd — Size of each Num. ber in decimal parts of an inch.	Old Stand'd — Difference between consecutive Nos. in dec. parts of an in.
0000	.460	.0504	.454	.029
000	.4096	.0448	.425	.045
00	.3648	.0399	.380	.040
0	.3249	.0356	.340	.040
1	.2893	.0313	.300	.016
2	.2580	.028	.284	.025
3	.2300	.025	.259	.021
4	.205	.0231	.238	.018
5	.182	.0199	.220	.015
6	.162	.0177	.203	.017
7	.1443	.0158	.180	.023
8	.1285	.0141	.165	.017
9	.1144	.0125	.148	.014
10	.1019	.01118	.134	.014
11	.09072	.00986	.120	.014
12	.08081	.00992	.109	.011
13	.07196	.00786	.095	.012
14	.06408	.00702	.083	.011
15	.05706	.00624	.072	.007
16	.05082	.00556	.065	.007
17	.04525	.00495	.058	.009
18	.0403	.00441	.049	.007
19	.03589	.0035	.042	.006
20	.03196	.00349	.035	.003
21	.02846	.00311	.032	.004
22	.02535	.00278	.028	.003
23	.02257	.0023	.025	.003
24	.0201	.00196	.022	.002
25	.0179	.00174	.020	.002
26	.0159	.00142	.018	.002
27	.01294	.00185	.016	.003
28	.0142	.00127	.014	.001
29	.01264	.0011	.013	.001
30	.01129	.00097	.012	.002
31	.01002	.00087	.010	.001
32	.00892	.00078	.009	.002
33	.00795	.00069	.007	.001
34	.00708	.00063	.006	.002
35	.00631	.00061	.005	.001
36	.005		.004	

THE STANDARD WIRE GAUGE,

Manufactured by

J. R. BROWN & SHARPE,

115 South Main St., Providence, R. I.

Adopted by the Brass Manufacturers, January, 1858.

These Gauges are made from the best steel, and are tempered, adjusted, and warranted accurate.

☞ None genuine unless stamped as in the engraving, with our trade marks.

PRICES.

Round Gauges, sizes, 0 to 36, . . . $2 50
"   "   (size of cut) 5 to 36, 2 00

A variety of **V** or angular Gauges on hand and made to order.

RESOLUTION OF THE BRASS MANUFACTURERS.

Whereas, it seems desirable that some steps be taken to arrive at a more complete uniformity in Wire Gauge, used by the Brass Makers, and whereas, J. R. Brown & Sharpe, of Providence, R. I., have, at considerable expense, prepared a Gauge with a new grade of sizes, a plan which is by us approved, therefore,

Resolved, That we will adopt said Guage, and be governed by it in rolling our metals, and will use our exertions to have it come into general use as the Standard U. S. Gauge.

Signed,

BENEDICT & BURNHAM MANF'G CO.,
   Chas. Benedict, Secretary.
BROWN & BROTHERS.
   Philo Brown, President.
SCOVILL MANF'G CO.,
   S. M. Buckingham, Treasurer.
THOMAS MANF'G CO.,
   S. Thomas, Jr., President.
NEW YORK & BROOKLYN BRASS CO ,
   John Davol, President.
NEW YORK BRASS & MANF'G CO.,
   J. Hoppock, President.
JAS. G. MOFFETT.

WATERBURY BRASS CO.,
   L. W. Coe, Treasurer.
HOLMES, BOOTH & HAYDENS,
   J. C. Booth, Secretary.
BRISTOL BRASS & CLOCK CO.
   E. N. Welch, President.
WALLACE & SONS,
   Thomas Wallace, President
ANSONIA BRASS & BATTERY CO.
   J. H. Bartholomew, Agent.
BELLVILLE WIRE WORKS,
   G. De Witt, Agent.
WALCOTTVILLE BRASS CO.,
   W. M. Hungerford, President.

Steel Squares, for Machinists.

Blades divided to 32ds of inches.

With 3 inch blade, . . . $1 40
"   4 "   "   . . . 1 50
"   6 "   "   . . . 2 00
"   9 "   "   . . . 2 50
"   12 "   "   . . . 3 50

PLUMB BOBS.

Brass, with screw cap and steel point, ... $1 25
Without screw cap 1 00
Iron, without screw cap, ... 50

STEEL STRAIGHT EDGES,
AND BLADES FOR **T** SQUARES, MADE TO ORDER.

A LIBERAL DISCOUNT TO DEALERS.

June, 1859.

BROWN & SHARPE MFG. CO., Providence, RI

The following 1889-1900 BROWN & SHARPE MFG. CO. ads expand on the information contained in the company's entry in *Makers of American Machinist's Tools,* first volume.

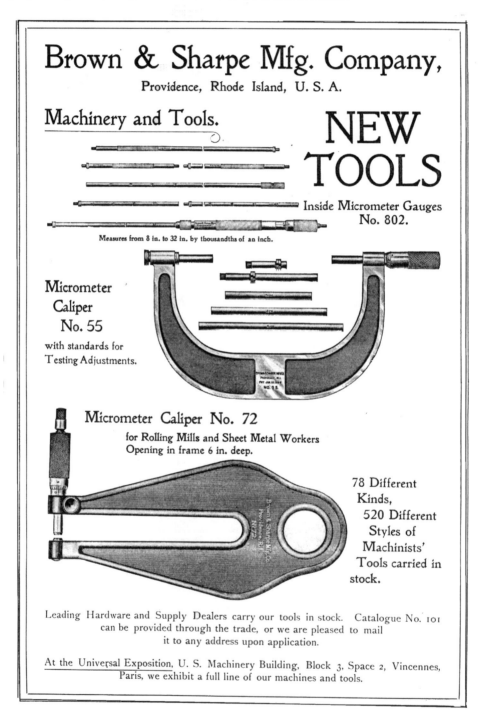

Machinery 1900

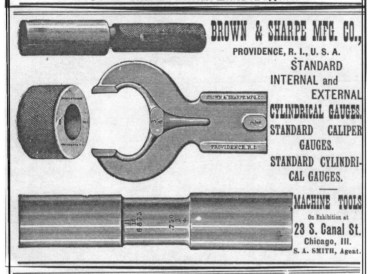

American Machiniest 1889

American Machiniest 1892

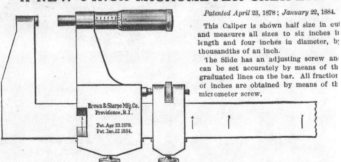

American Machinist 1893

BUSH, T. DANIEL, South Allentown, PA

Maker, in 1919, of a "universal indicator" which is actually a wiggler as shown below. Bush also furnished an attachment to mount the tool on a surface gage. Later in 1919 he began offering another attachment which allowed the tool to be used as a lathe center indicator.

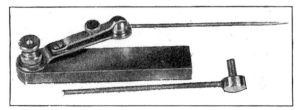

BUSH UNIVERSAL INDICATOR

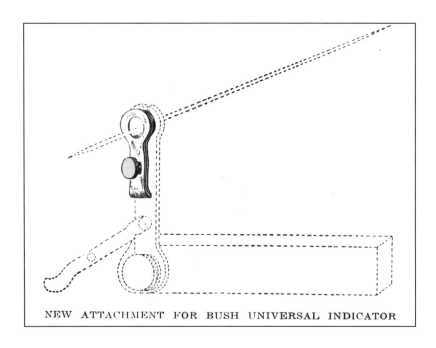

NEW ATTACHMENT FOR BUSH UNIVERSAL INDICATOR

CALL, A., Springfield, MA

Operated by Amos Call (1814-1888). Maker of a combined gage and caliper nearly identical to those made later by DARLING & SCHWARTZ and later still by DARLING, BROWN & SHARPE. Call also made machinist's bevels and small machinist's squares. A. CALL tools are seldom encountered.

Call's history is a bit uncertain. Apprenticed to Stephen C. Bemis in 1828, he became a partner in BEMIS & CALL, Springfield, MA, when it was formed in 1844. Call was elected an officer when the firm reorganized as the BEMIS & CALL CO. in 1855 and was president of the firm when he died in 1888.

Tools marked with his name appear to have been made during the 1828-1844 period when he was supposed to have been working for Bemis. It seems likely that Call left to operate his own business for a short period and later rejoined Bemis. It's also possible that both Bemis and Call made tools under their own names during the partnership.

Call tools, such as shown below, will also be found marked BEMIS & CALL or BEMIS & CALL CO.

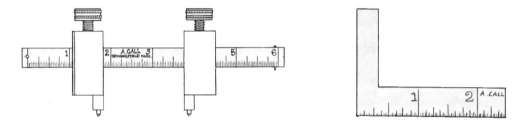

CHAMBERLIN, CHARLES W., Worcester, MA

Chamberlin had been a partner in the caliper and divider making firm of COPELAND & CHAMBERLIN beginning about 1874. By 1881, he was operating on his own and advertised "Dividers, Calipers, and Window Springs. Jobber of Light Machinist's Tools". This advertisement was illustrated with a cut of the Cooke's patent divider which had been made by COPELAND & CHAMBERLIN.

CHAMBERLIN, G.L., Pittsburgh, PA

Inventor of a combination level, try-square, clinometer, bevel protractor, etc., patented January 1, 1867. The tool was advertised as "suitable for the machinist, pattern maker, and draftsman."

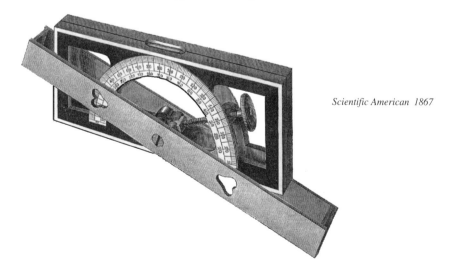

Scientific American 1867

CHANDLER & CO., RUFUS, Springfield, MA

Advertised in 1864 as "manufacturers of all kinds of machine screws, also Steel Slide Gages, other mechanics tools, and small machinery generally." As noted in *Makers of American Machinist's Tools*, first volume, the company was reported as making vernier calipers about 1880. Only a very few are known to exist.

CHICOPEE FALLS CO., Chicopee, MA

Maker of a round wire gage similar to the BROWN & SHARPE MFG. CO. type.

COES, CHARLES   See MICROMETER CALIPER CO.

COLUMBIA CALIPER CO., Marietta, PA

Successor to E.G. SMITH, Columbia, PA, probably formed soon after Smith moved to Tampa, FL, in 1915. Catalogs are known with a COLUMBIA CALIPER CO., MARIETTA, PENNA., label pasted over the E.G. SMITH name and address. A second label is pasted on the back cover stating: E.G. SMITH, REMOVED TO 315 W. PARK AVE., TAMPA, FLA. The contents indicate that the COLUMBIA CALIPER CO. took over the complete line of slide calipers, spherometers, micrometers and rules offered by E.G. SMITH. As noted in the E.G. SMITH entry, the tools appear to be of European origin and may have been imported rather than made by Smith or Columbia.

COMBINATION TOOL CO., New York, NY

Maker of a combination tool patented November 23, 1869, by Joel Manchester of New York City. The only known example of this tool is identical to the advertising cut below. It barely qualifies as a machinist tool but was found with a group of ca.1855-1870 machinist's tools all owned by the same machinist.

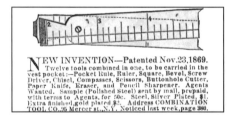

Scientific American 1869

COOK, J.H., Syracuse, NY

Maker of inside micrometer sets, machinist's trammels, and attachments for converting standard rules into hook rules, ca.1885-1902. The inside micrometers are marked only with his name and city. The hook rule attachments are marked PAT APPLD FOR, a model number, and Cook's name and city. The attachments must match the width and thickness of the rule as well as the length, so the model number must indicate the rule which fits the attachment. Two specimens have been observed: NO. 3.1 which fits a D. B. & S. 12" rule and NO. 11¼ which fits a D.& S. 12" rule.

Cook has proven to be something of a mystery. He is the same J.H. COOK listed in *Makers of American Machinist's Tools,* first volume, as the maker of a surface gage in Millers Falls, MA, (identical surface gages marked with the Syracuse address have been noted) and the same Cook associated with both the L.S. STARRETT CO. and SAWYER TOOL MFG. CO.

It is clear that inside micrometers and machinist's trammels introduced by Sawyer about 1900 are identical to those marked J.H. COOK, SYRACUSE N.Y. We also know that when Cook joined Sawyer in 1903 the firm announced that: "J.H. Cook, who for over 20 years has been connected with the L.S. Starrett Co., Athol, Mass., in the sales department, has resigned to accept a similiar position with the Sawyer Tool Mfg. Co. of Fitchburg, Mass." If accurate, this statement would mean that Cook operated his own tool business and sold or licensed designs to Sawyer, while a Starrett employee or contractor.

It also suggests that J.H. Cook is the designer of the Cook's improved trammel points listed in Starrett catalogs as early as 1887, and Cook's extension beam trammels listed by 1888.

The illustrations below show the trammels as offered in the Machinist's Supply Co. catalog of 1891. Although not specifically stated to be Starrett products, they appear with other Starrett tools. It is quite possible that Cook at first made the trammels and sold them through Starrett. No "Starrett" style trammels marked with Cook's name have been reported, however.

Despite the PAT APLD FOR markings on the surface gage and hook rule attachments and references to "Cook's Patent" trammels, no Cook patents are known except his August 22, 1905, caliper joint patent assigned to the SAWYER TOOL MFG. CO.

Perhaps the strangest circumstance is the lack of any reference to J.H. COOK tools in any comtemporary magazine or catalog reviewed. Cook must have sold his products through some commercial channel, but that channel is still unknown. *(Illustrations continued on next page.)*

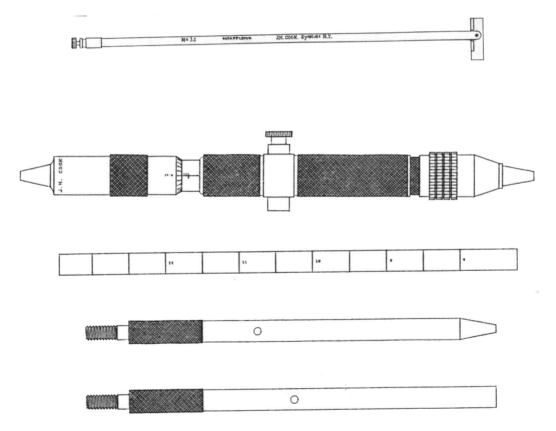

COOK'S IMPROVED TRAMMEL POINTS.

Made of Bronze Metal, with Forged Steel Point, Hardened.
Either point can be removed, and the pencil socket accompanying each pair, put in its place.
The best Trammel Points in the market. Adjustable like *spring dividers*. Light and durable. For bar ⅝ x ⅜.

PRICE:
With 3 inch points, adjustable......... $2 50
With 3 inch points, not adjustable..... 1 50
Extra long points, 5 inch, per set 55

COOK'S NEW EXTENSION BEAM TRAMMELS.

This cut represents a pair of trammel heads with an opening through the under side to accommodate the extension, giving width and stiffness in proportion to the length required for large work, while it is equally well adapted to receive a narrow beam for light work.
The points are eccentric, and may be loosened and rotated in their sockets to make fine adjustments. Either point may be removed and a common pencil inserted.
The marks on legs enable them to be adjusted in proper relation to each other.

Price, complete $3 25
" without caliper legs, 2 50

COOKE & CO., A.A.& G.L., Worcester, MA

Maker of Cooke's patent dividers, patented December 12, 1871, by Albert A. Cooke. The firm appears to have been active from about 1872 until it sold out to Salem Copeland about 1874. The below advertisement appeared in the 1873 Worcester City Directory.

(See illustration next page.)

COPELAND & CHAMBERLIN, Worcester, MA

Makers of the Cooke's patent dividers beginning about 1874 when they bought the line from A.A. & G.L. COOKE & CO. See entry in *Makers of American Machinist's Tools,* first volume, for other information.

The firm was also the maker of the Wight patent slide caliper, patented by Hiram P. Wight, Worcester, MA, on June 25, 1878. Wight was listed as the maker in *Makers of American Machinist's Tools,* first volume, but existing specimens are marked:

COPELAND & CHAMBERLIN
WORCESTER, MASS.
PATENTED JUNE 25, 1878

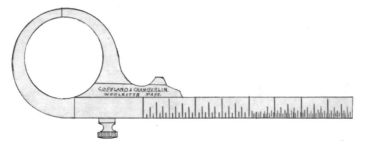

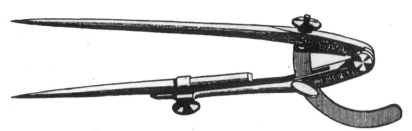

No. 3.

COOKE'S PATENT
EXTENSION DIVIDERS.

Improved Patent Attachment of Movable Point.

PATENTED DECEMBER 12, 1871.

MANUFACTURED BY

A. A. & G. L. COOKE & CO.,

WORCESTER, MASS.

ORDERS FILLED BY

S. W. COOKE, 442 MAIN ST.,

Of whom Price-Lists and Circulars may be obtained.

THESE DIVIDERS will commend themselves at sight to every practical mechanic. They answer all the purposes of the common Dividers; and the MOVABLE POINT renders them of advantage, many times, over the old style.

A Pencil can be used in place of the Movable Point,

and they are thus made to answer all the purposes of DRAFTING DIVIDERS.

Draughtsmen, Joiners and Carpenters,

and all who have occasion to use Dividers will find these to excel in utility, the old style.

Six different sizes are made, all having the pencil leg. The larger sizes are particularly adapted to meet the wants of wheelwrights, carriage makers, stair builders, bonnet block or pattern makers, or any work requiring a large sweep or size, or unusual length of legs. Caliper points are made for 10 in.—15 in.

They may be obtained of Hardware Dealers throughout the country.

No. 3   No. 4.

Represents the 6, 7, 8 and 10 inch sizes, with one movable leg that can be lengthened or shortened, as need requires; or can be substituted for the pencil when desired.

The Attachment that holds the leg or the pencil, secures the leg firmly, is easily adjusted, and is readily used.

No. 4

Represents the 15 and 10 inch sizes, each with double legs or movable points. The 10 inch becomes 15 inch, and the 15 inch becomes 20 inch, merely by the adjustment of the legs.

CORBIN SCREW CO., New Britain, CT

Maker, from about 1914, of a speed indicator which was a true tachometer, reading directly in RPM. The line was taken over by the VEEDER-ROOT CO., Hartford, CT, in 1929.

CRESCENT

Name cast into the frame of the micrometer shown below. It has a very early appearance but has no indentifying marks to show maker or time period.

CUT INDICATOR CO., New York, NY

Maker, in 1906, of a cutting speed calculator which converted part (or tool) diameter and RPM into surface speed in feet per minute. This figure became necessary when high speed tool steel came into widespread use about this time. Many such devices were offered, but this one, with a nickel case and glass face is certainly one of the most interesting. The basic device, in cardboard or plastic, remains a popular "giveaway" by machine tool and cutting tool makers.

American Machinist 1906

THE CUT INDICATOR.

DANIELS, G.W., Waltham, MA

Operated by George W. Daniels (1830-1886), maker of wing calipers and dividers, and small machines for the watch making industry.

DARLING, EDWIN A., Hartford, CT

Inventor of a surface gage patented April 17, 1883. One example has surfaced, marked only with the patent date. Darling may or may not have been the maker.

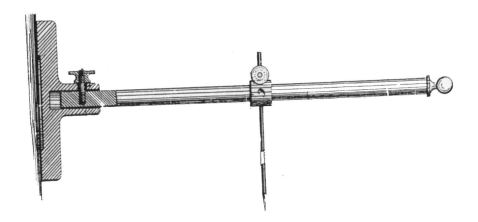

DARLING, SAMUEL, Providence, RI

It's well known that Samuel Darling was a member of the firm of DARLING, BROWN & SHARPE from 1866 until selling out to his partners on January 1, 1892. However, an odd, unmarked micrometer, identical to his patent for a Micrometer Gauge for Wire and Sheet Metal, granted December 7, 1880, is in the Smithsonian collection. Darling may have made these outside of the DARLING, BROWN & SHARPE partnership.

Photo courtesy of the Smithsonian Institution, file number 308976.

DETRICK, Jacob S., San Francisco, CA

Granted patent for a "counting register" August 4, 1868. This is the earliest speed indicator patent showing the worm and gear wheel mechanism in common use through much of the 19th and early 20th centuries. Detrick, however, claimed only an improvement in the mechanism and implied that the basic design was well known in 1868. No speed indicators marked with Detrick's name have been observed, but unmarked indicators with his patented feature (two wheels, one with 100 teeth, the other with 101 teeth) are known.

By 1882, Detrick was building machinist's planers in Baltimore, MD, and formed the partnership of Detrick & Harvey in 1884.

DONALDSON, JOHN, St. Louis, MO

Inventor and maker of an adjustable key-way gage, patented November 19, 1872.

Scientific American 1873

DORDEA MFG. CO., Detroit, MI

Maker, about 1920, of fine adjusting machinist's beam trammels.

DRAKE & CO., Lake Village, NH

Possible maker of a thread gage identical to those made by the DAVIS LEVEL & TOOL CO. The known example is marked only DRAKE & CO., LAKE VILLAGE, N.H. It's quite possible that this is an advertising piece and not actually made by Drake.

DUFFY, ANTHONY, Poughkeepsie, NY

Maker, in 1904, of a machinist's tool which combined an adjustable square, center square, caliper, height gage, and adjustable straight edge. Duffy was granted a patent for the tool on September 8, 1903. As was the case with many combination tools, it was not a commercial success.

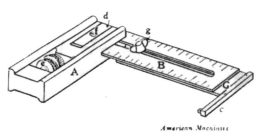

American Machinist 1904

MACHINIST'S TRY-SQUARE.

DUNHAM, C.N., Philadelphia, PA
Maker of machinist's calipers patented July 29, 1884, by William P. Dadson.

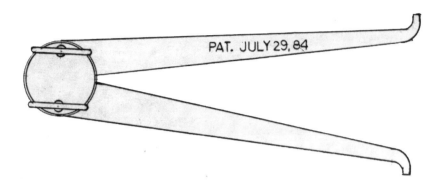

E.N.W.Co. See ELGIN NATIONAL WATCH CO.

ELGIN NATIONAL WATCH CO., Elgin, IL
An unusual style of surface gage marked E.N.W.Co., was made by this company. They were, of course, a watch making company from formation in 1864, but appear to have also made some tools. The surface gage may have been made for internal use only, but it is clearly a factory designed-and-made tool of high quality.

EMPIRE MACHINE CO., Pittsburgh, PA

Maker, in 1905, of an unique speed indicator. The rotation of the input shaft caused an internal fan to rotate which, in turn, caused a vane to turn against a hairspring an amount porportional to the velocity of the air current created by the fan. Note from the sectional view below that the tool was quite complex and probably rather fragile. The indicators were made in two sizes; one reading 50 to 400 RPM and the other 300 to 2500 RPM. No examples have been observed.

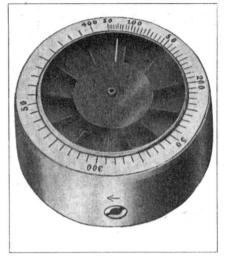

American Machinist 1905

FIG. 1. A TACHOMETER.

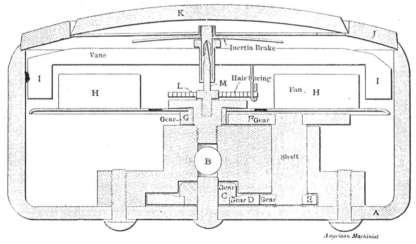

FIG. 2. SECTIONAL VIEW OF A TACHOMETER.

ERB MFG. CO., E.M., Jersey City, NJ

Maker of machinist's dial test indicators patented October 24, 1911, by Edmund M. Erb. The company noted that the face was covered with celluloid instead of the "usual watch crystal" and that the spindle was made with a shoulder that prevented overtravel into the mechanism. The below detail drawing was published in *American Machinist* magazine April 20, 1911. *(See next page for illustration.)*

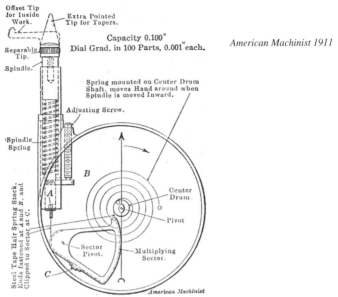

American Machinist 1911

FIG. 2. DETAILS OF DIAL TEST INDICATOR

FANEUF & TURNER, Orange, MA

Maker of machinist's try squares. See F.L. TURNER & CO. and GREBLE, TURNER & CO.

FINDER MFG. COMPANY, St. Paul, MN

Maker of the FINDER indicator, probably in the 1920s or 1930s. The tool was made of sheet metal stampings, and as shown below, sold for only $1. It is very similar, in both form and function, to an English-made indicator patented in the U.S. on February 23, 1915, and marked only UNIQUE. *(See illustration on next page.)*

FINLEY CO., J.A., Allston, MA

Maker, in 1922, of the "Duwell" self-guiding scriber for toolmakers. The tool had a double spring point on one end "by means of which the operator is enabled to lay out an outline at the bottom of a small, irregular-shaped hole." They also offered the "Duwell" centering device which located and punched the center of round stock.

FISHER-SMITH CO., Dayton, OH

Maker, in 1919, of a Ball Test Arbor (wiggler) for "truing work with the spindle of a machine." In looks and function it is almost identical to several other such tools.

FITCHBURG TOOL CO., Fitchburg, MA

Name found on a machinist's combination square. No such company has been located, so the name may be a trade name used by some other company, such as the SAWYER TOOL MFG. CO., Fitchburg, MA.

Illustration from the Finder Mfg. Company entry, preceding page.

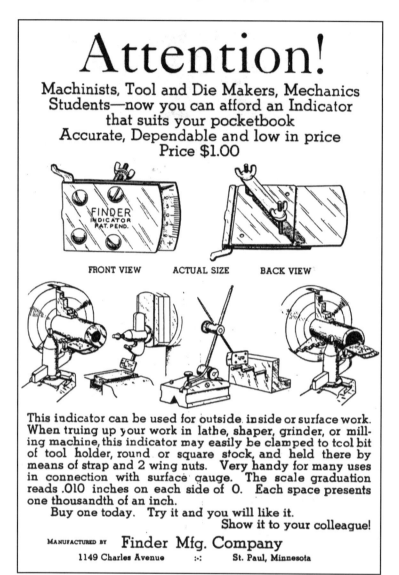

FLORENCE IRON WORKS, Florence, CO

Maker, in 1910, of a layout protractor patented February 8, 1910, by William E. Perkins, company vice president. "The instrument will be found particularly useful in laying out a line of keyways along shafts, either all in line or at specified angles with each other." The central rod could be used as either a scriber or center punch.

(See next page for illustration.)

American Machinist 1910

A LAYING-OUT PROTRACTOR

FOOT & CO., H., address unknown

The combination caliper and divider shown below is marked H. FOOT & CO., PATENTED on one leg. The other leg is marked with the user's name, H.K. SMITH, and the date 1852. No patent related to this tool has been found.

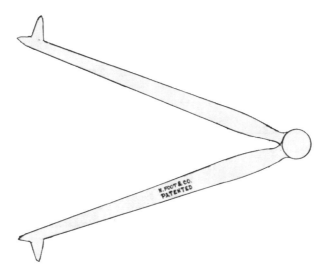

FOOT, JAMES D., New York, NY

Foot was shown as a probable dealer in *Makers of American Machinist's Tools,* first volume, It now appears that he was, in fact, a manufacturer of machinist's tools until 1881. The following notice appeared in the *American Machinist* magazine in January, 1881:

"James D. Foot, 78 Chambers street, New York, formerly interested in the manufacture of small machinist's tools, has sold out his interest to C.L. Bellamy (the orginator of these tools), his file business having become so extended as to force him to give up all outside lines of manufacture."

FORBES & CO., W.D., New London, CT

Operated by William D. Forbes (1852-1921). Forbes founded the company about 1887 to make milling machines in Hoboken, NJ. He moved to New London, CT, in 1908 where he concentrated on making small steam engines. In 1915, he introduced the surface gage shown below. No examples have been examined.

A somewhat novel form of surface gage has recently been brought out by W. D. Forbes, New London, Conn. It is very simple in construction, which enables it to be sold at a low price, and yet has a number of interesting features to commend it. Rapid vertical adjustment is obtained by sliding the arm on the column, and by means of a thumbscrew it may be shifted into any position desired. The thumbscrew bears against a copper washer, to prevent bruising the column. A coarse vertical adjustment is obtained by tilting the beam on the arm, the thumbnut holding it securely in place. Fine or micrometer adjustment of the needle is obtained by the knurled nut at the right. The needle being held at two points makes it very stiff, and is not liable to move out of adjustment. The base is $2\frac{1}{4}$ in. square by $\frac{3}{4}$ in. thick and is drilled with two $\frac{1}{8}$-in. holes in which plugs can be placed when the cage is used to set work parallel to the edge of a planer-head or similar machine, or where it is more convenient to use the gage on the planer rail than from the face of its platen. The gage is $9\frac{1}{4}$ in. high; the needle is 9 in. long. The total weight, $1\frac{3}{4}$ lb.

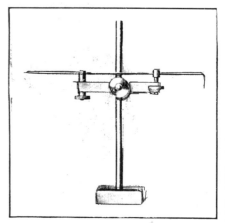

SURFACE GAGE

American Machinist 1915

FORREST, B.L., Fitchburg, MA, and
FORREST MFG. CO., B.L., Fitchburg, MA

The entry in *Makers of American Machinist's Tools*, first volume, states that B.L. Forrest was a maker of steel machinist's rules. It now appears, however, that the rules were made by the SAWYER TOOL MFG. CO. using B.L. FORREST and B.L. FORREST MFG. CO. as trade names. Bert L. Forrest was first listed in the Fitchburg City Directory in 1908 and was shown as "emp Sawyer Mfg. Co." No listing appears for the B.L. FORREST MFG. CO. Forrest's 1912 listing, stating that he had moved to Leominster, coincides with the move of the SAWYER TOOL MFG. CO. to Ashburnham, MA.

Many Forrest rules have been found in leather sleeves printed with advertising. It would appear that the Forrest line was a low cost one made for sale to companies which gave them to customers.

FULTON

Trade name used by MONTGOMERY & CO., New York, NY.

GENALES, ANDREW, Nyack, NY

Apparently the last maker of the Henry Koch patent (July 17, 1906) test indicator that had previously been made by H. KOCH, and later, H. KOCH & SON. The below Genales flyer also shows the "Positive Lock" surface gage and machinist's surface base which he offered. The flyer is undated but the seven digit telephone number would indicate post-WWII.

NEW IMPROVEMENT

Our Positive Lock Spindle Surface Gauge that can not slip and loose your setting.

Our Hardened Bases that can not scratch.

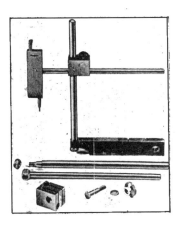

The set of Attachments with the Koch Test Indicator and tool post holder makes it reading in all positions. Adaptable to Lathe, Shaper, Jigbora, Milling Machine Drill Press and many other uses.

Spindle 6" long 5/16" diameter.
Indicator extension Rod 6/ long.
Face of Indicator can be turned to any position.

RETAIL PRICES

Koch Test Indicator
with tool post holder $10.00

1 Koch Indicator head, stud and nut 9.00

1 Koch Lathe Tool Post holder
⅝ x ⅜ x 4½ 1.00

1 Spindle 5/16 D. 6" long and nut to fit Koch Indicator holder 1.00

Koch Indicators can also be furnished in Metric reading 1/50 of a millimeter, 12 lines plus 6 minus 6 at same price.

1 Surface gauge attachment for 5/16" spindle with 6" extension rod and Clamp 4.00

1 Extra unit, clamp for 5/16" spindle and scriber 4.00

1 #1 Universal Tool Makers Surface Gauge with Positive Spindle Lock and scriber. Spindle 5/16" D. 9" long. Base 3¾ x 2½ x1.
with
HARDENED BASE $9.75
SOFT BASE $8.75

1 #2 Manufacturers and Machinists Surface Base 3 x 2 x 1 with Spindle 6" long 5/16" D.
with
HARDENED BASE $5.30
SOFT BASE $4.30

Plus Postage
Discount to Dealers Only
Subject To Change Without Notice

Additional illustration on next page.

THE KOCH TEST INDICATOR
IS THE ONLY TOOL MAKERS INDICATOR ON THE MARKET WITH TWO WORKING ENDS, ONE END TO TEST OUTSIDE SURFACES, THE OTHER INSIDE.

The KOCH TEST INDICATOR is scientifically designed to give extreme sensitiveness combined with ruggedness. The soft, smooth action of the KOCH TEST INDICATOR added to its high magnifying power make it the choice of master toolmakers, machinists, and inspectors.

It is the only indicator with low friction that has a torsion spring as well as a compression spring, assuring positive and sustained accuracy.

Each graduation on the scale of the KOCH TEST INDICATOR represent 1/1000 of an inch movement of the plunger.

The KOCH TEST INDICATOR is so designed that the plunger moves away from instead of against the lever preventing the delicate parts from being broken by a sudden or excessive jolt of the plunger.

The case of the indicator is made of cast iron which will not warp or scratch. The details are made of tool steel and the working contacts and bearings are hardened to resist wear.

The KOCH TEST INDICATOR can now be obtained with a complete set of attachments which make it adapted for inspection work on a surface plate, setting up work on milling machines, jigborer and testing flat surfaces where surface gauges are frequently used.

Koch Indicators can also be furnished in Metric reading 1/50 of a millimeter, 12 lines plus 6 minus at same price.

This shows various uses of the Surface Gauge and Indicator attachment to test Inside and Out-side surfaces with the Koch Test Indicator. Also the Indicator with lathe tool post holder.

New improvement on Gauge #1 Safety feature that Positive Lock Spindle that can not slip and lose your measurements. Toolmakers universal Surface Gauge with scriber. The spindle draw bolt has a positive lock and when set to measurements for height or drawn parallel lines you can be assured that the gauge will not slip.

Size of Gauge Spindle 9" long 5/16 Diameter Block 3¼ x 2½ x 1 and has a 90 degree angle at base to fit any surface of 90 degree angle as well as to be used on cylindrical work. The Surface of gauge is precision ground. Made of Chrome casting that will not scratch so easy (The soft base). Also furnished in hardened bases. The Fine adjustments of the gauge and positive lock makes it the perfect tool for laying out work of precision on Jigborer, milling machine and inspection plate. Manufacturer Andrew Genales, Koch Test Indicator, Nyack, New York.

Telephone NYack 7-2222

GILBERT, HARRIS & CO., Chicago, IL

Maker, in 1910, of universal micrometer calipers patented August 9, 1910, by Herman Sauter. Sauter had assigned the patent to this company. Examples of both styles shown below have been found in the Chicago area.

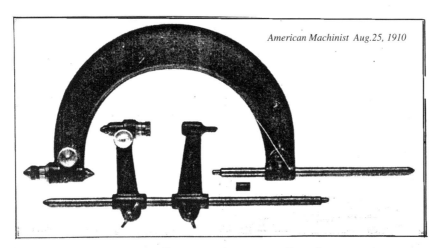

FIG. 1. UNIVERSAL MICROMETER CALIPER

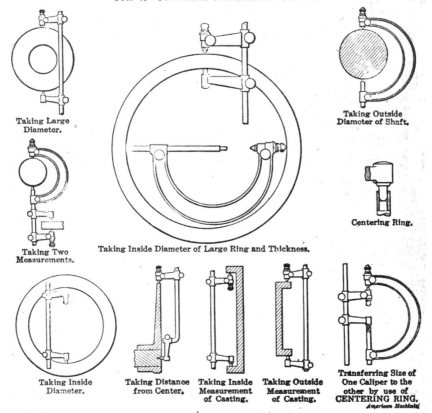

FIG. 2. SOME APPLICATIONS OF THE UNIVERSAL MICROMETER CALIPER

GLADWIN, Flint, MI

Maker of test indicators patented March 19, 1935, and June 28, 1938, by Alfred G. Winters.

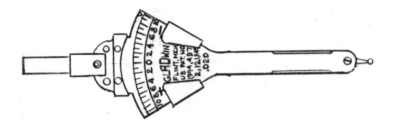

GLEASON, E.P., New York, NY

Maker of an adjustable tap wrench, patented June 19, 1860.

GOULD, FAYETTE, Huntington, L.I.

Inventor of dial calipers patented April 12, 1859. The calipers operated on the rack and pinion principle and are identical, in function, to calipers made now and thought by present day machinists to be a recent improvement on the vernier caliper.

Scientific American 1859

GOULD'S CALLIPERS.

b, sunk into its upper edge. The movable leg of the calliper, C, slides on A by a groove, the lower side of which, a, is in contact with the lower or even side of A; the upper part of C has an arbor, d, through it, that, at the back, Fig. 2, carries a toothed wheel, c, that fits in the rack, and in the front, Fig. 1, a pointer or index, e. To the front of C a disk, D, is fixed, having graduations, f, all around it, dividing it into any number of inches and parts of an inch, with relation to the inches on A, as may be desirable, so that, as C is moved the toothed wheel, c, rotates, carrying with it the pointer, e, by whose means the exact size of the object or the distance apart of the callipers may be accurately noted.

The inventor, Fayette Gould, of Huntington, L. I., will give any further information desired upon application to him. The patent is dated April 12, 1859.

GREBLE, TURNER & CO., Hamilton, OH, and Orange, MA

See listing in *Makers of American Machinist's Tools*, first volume, and F.L. TURNER & CO. and FANEUF & TURNER in this volume.

GREELEY & CO., E.S., New York, NY

Maker, ca.1884-1888, of the Pratt Speed Indicator, patented June 3, 1884, by Robert J. Pratt. The 1887 offering in the Palmer, Cunningham & Co. catalog is shown below.

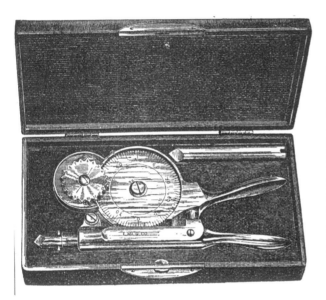

In this Speed Indicator the toothed dial is carried by an arm hinged to the main frame, and within the frame is journaled a loosely-turning worm-spindle. The main frame and arm are provided with hand grasps. When in operation the spindle is turned loosely in its bearings by the shaft. The operator by means of the handle of the pivoted arm can instantly gear the indicating dial with the worm shaft at the very beginning of the period of time for which the velocity is to be determined. At the end of each period the dial is as quickly disengaged—the recoil of a spring, when the handle is released, forcing the pivoted arm at once from the spindle. A set screw governs the approach of the dial-teeth to the worm, and prevents the dial-teeth from being forced or crowded in upon the spindle with so much friction as to abrade or wear off their faces. The instrument is finished in first-class style, and can be used for right or left motion.

Nickeled Indicator, with extra second dial and tip, in case, complete, (as shown in cut), $6.00.

GREENE, TWEED & CO., New York, NY

Maker of the L.T. Weiss patent (November 22, 1892) speed indicator after the first maker, WEISS BROTHERS, went out of business. In 1938, the firm offered the Empire speed indictor shown below. Note the similarity to the Brown & Sharpe No. 746 speed indicator.

Machinery 1938

GREGORY, H.R., Plattsburgh, NY

Agent for the Wilkinson combination tool patented October, 27, 1868, by John D. Wilkinson and E.O Boyle of Plattsburgh, NY. The tool combines a wire gage, rule, square, inside calipers, outside calipers, dividers and center gage which folds into a 2" diameter. The only known example is identical to the 1871 advertising cut below except that it does not include the wire gage. See also WILKINSON & CO. in *Makers of American Machinist's Tools,* first volume.

Scientific American 1871

CAN be carried in the vest pocket, and combines seven full-sized tools, indispensable to mechanics. Price, $3.00. H. R. GREGORY, General Agent, Plattsburgh, N. Y.

GRIMSHAW & BAXTER, England

Although not an American machinist's tool, the "Equating" micrometer, shown below, is included because of its unique design and its distribution in the U.S. The tool is a micrometer with a worm gear cut into the barrel, turning a gear wheel which is calibrated to display equivalents of the micrometer opening in other measurements (stubs wire gage in the example shown). It was patented in the U.S., October 9, 1909, by H. Brooke and A.J. Baxter. Several examples are known.

Machinery 1903

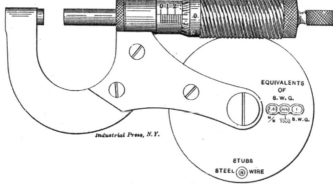

Fig. 4. English Equating Micrometer.

GULDAGER & JANTCH CO., Detroit, MI

Maker, in 1920, of the "Correct" indicator holder designed to be mounted on a machinist's square.

H. & S. , Lowell, MA See HUNTOON & SIMONDS
H.S. & CO., NY See HAMMACHER, SCHLEMMER & CO.

HAINES GAUGE CO., Philadelphia, PA.

Maker, in 1893, of the Haines Automatic Micrometer Rolling Mill Gauge patented February 9, 1892, by Robert Haines, Jr. The gauge, shown below, was designed to allow accurate thickness measurement (to .0025") of very hot material, such as is found in a steel rolling mill. A spiral spring caused the micrometer head to rotate when the trigger, located several inches away from the micrometer head, was operated. The measurement therefore, could be taken without close exposure to the hot metal. The gauge must have been popular with rolling mill men, as examples are fairly common today.

American Manufacturer 1893

HAMMACHER, SCHLEMMER & CO., New York, NY

Formed in 1885 by Albert Hammacher and William S. Schlemmer. A large New York hardware house, the firm offered a number of tools made by it (or for it) under its own name. Such tools were usually marked H.S. & CO. *Its machinist's tool case is shown on the next page.*

HANSEN, G.L., Rockford, IL

Inventor and maker of a machinist's test indicator patented May 8, 1906. The tool is extremely well made and finished and was likely expensive to make. The only known example is marked:

G.L. HANSEN, MFR.
ROCKFORD, ILL.
PAT. MAY 8, 1906

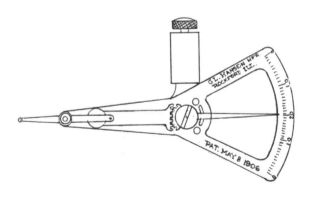

Machinery 1910

THE H. S. & CO. MACHINISTS' PORTABLE TOOL CASE

EVERY INCH OF SPACE AVAILABLE!

Showing case closed and locked.

Showing case open and drawers exposed.

This is the only practical *portable* tool case on the market. The drawers are so arranged that every available inch can be utilized, and the tools so distributed that each one can quickly and easily be picked out and without first handling over a number of other tools which have hurriedly been placed on top of the one wanted. With the exception of the bottoms and sides of drawers, it is made throughout of thoroughly kiln-dried solid (not veneered) oak, which will not warp nor twist, and very carefully and handsomely polished and finished; the corners are dovetailed and the workmanship is of the very best. The trimmings are of polished brass; has a brass spring-lock with two flat steel keys, and a comfortable leather handle for convenience in carrying.

The *outside* measurements of the case when closed are: $15\frac{5}{8}$ in. long, 8 in. deep, $10\frac{3}{4}$ in. high. The *inside* dimensions of the eight drawers are as follows:

	long	deep	high
First	$6\frac{1}{2}$ in.	$6\frac{1}{8}$ in.	$\frac{3}{4}$ in.
Second	$6\frac{1}{2}$ "	$6\frac{1}{8}$ "	$2\frac{3}{16}$ "
Third	$6\frac{1}{2}$ "	$6\frac{1}{8}$ "	$\frac{3}{4}$ "
Fourth	$6\frac{1}{2}$ "	$6\frac{1}{8}$ "	1 "
Fifth	$6\frac{1}{2}$ "	$6\frac{1}{8}$ "	$1\frac{5}{8}$ "
Sixth	$14\frac{1}{8}$ "	$6\frac{1}{8}$ "	$1\frac{1}{2}$ "
Seventh	$14\frac{1}{8}$ "	$6\frac{1}{8}$ "	1 "
Eighth	$14\frac{1}{8}$ "	$6\frac{1}{8}$ "	$3\frac{3}{16}$ "

The bottoms of drawers are felt-lined, run on hardwood slides, and have brass polished knobs.

Machinists will find the little mirror which is placed on the inside of the case a very handy and convenient addition.

The weight is $14\frac{1}{2}$ pounds—packed for shipment approximately 30 pounds.

$10.00 net f.o.b. New York.

We can also furnish this case in mahogany finish at the same price.

HAMMACHER, SCHLEMMER & CO.
Hardware, Tools and Supplies

4th Ave. and 13th St.   NEW YORK, SINCE 1848

HAZELTON, FREDERICK D., Boston, MA

Hazelton was erroneously listed in *Makers of American Machinst's Tools,* first volume, as the maker of "Hazelton's Improved Caliper Attachments." He was the patentee, on November 21, 1876, but the tool was made by STONE & HAZELTON, a partnership which included Hazelton.

HENNECKE-WALKER CO., Milwaukee, WI

Formed in 1902 by the merger of the WALKER TOOL CO. and the F.J. Hennecke Machine Works, both of Milwaukee, WI. The new company continued production of the Walker caliper attachments and lathe center grinders.

Machinery 1902

HENRIKSON, ADOLPH F., Troy, NY

Maker, in 1906, of a test indicator designed to "test alignment of lathe centers, the truth (squareness to axis of rotation) of face plates and chucks, for setting work accurately in the miller and under drill spindles and for many other purposes." The linkage causes a sleeve (B in the drawing below) to move with any misalignment of the parts tested. The tube under the sleeve was calibrated in thousands, thus giving a direct reading of the error. The tool appears both useful and practical, but none have been observed.

Henrikson also received patents for a micrometer adjustment for linear scales on July 31, 1906, and calipers and dividers on May 26, 1908. No tools made under either patent have been observed.

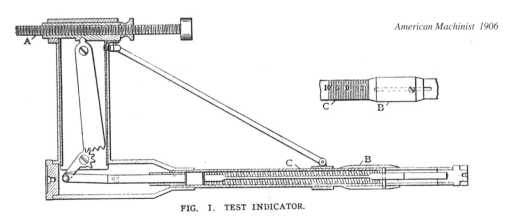

American Machinist 1906

FIG. I. TEST INDICATOR.

HEYDRICH, ADOLPH, Topeka, KS

Inventor and probably maker of an inside micrometer patented April 6, 1886. Topeka, KS, seems a strange place for a machinist's tool maker, but it was the site of a large railroad shop at the time. The only known example with a brass body, shown below, is marked only, PATD APR 6, 1886.

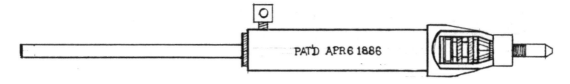

HILL MFG. CO., M.B., Worcester, MA

Maker, in 1914, of toolmaker's clamps and a universal test indicator, both introduced that year. In 1921, they were advertising two styles of test indicators. The newer style appears indentical to indicators previously made by the C.E. ROBINSON CO., Orange, MA. Hill probably took over the Robinson product, adding it to his 1914 test indicator line.

American Machinist 1921

Fig. 2. Hill Universal Test Indicator

Machinery 1914

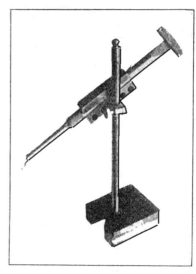

HILL TEST INDICATOR

HJORTH LATHE & TOOL CO., Boston, MA

Founded in 1912 by Henry J. Hjorth when he took over the shop and products of the REMINGTON TOOL & MACHINE CO. where he had been part owner. Twist drill gages, made by Hjorth under his patent of February 29, 1916, are nearly identical to those offered by Remington in 1909. (This seems strange until we note that Hjorth's patent application was made on March 31, 1909.) Micrometer adjusting depth gages previously made by Remington were also continued by Hjorth.

We now know that the micrometer surface gage offered by the REMINGTON TOOL & MACHINE CO. in its 1909 catalog was the Hjorth gage made under his patent of May 31, 1904, and later made by the HJORTH LATHE & TOOL CO. See entry in *Makers of American Machinist's Tools,* first volume, for illustrations and additional information.

HOGGSON & PETTIS MFG. CO., New Haven, CT

The below pages from the company's 1901 catalog include a previously unrecognized tool, Carr's Universal Scale Square. It's interesting to note that the scales making up the square are from COFFIN & LEIGHTON and that the complete line of COFFIN & LEIGHTON scales was listed elsewhere in the catalog. See entries for JAMES CARR, COFFIN & LEIGHTON, and HOGGSON & PETTIS MFG. CO. in *Makers of American Machinist's Tools,* first volume, for more information.

Carr's Universal Scale Square.

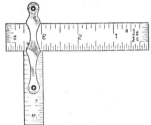

Fig. 46

A handy combination for laying out work. Can be used as a depth gauge, and the clamp lever on the two-inch scale can be used as a handle to facilitate measurements inside dies, and work difficult to get at.
Price 90 cents. Code word, *Usage.*

CARR'S PATENT
Improved Combination Scribing Block.

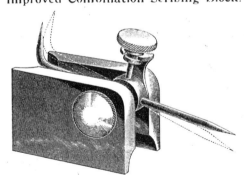

Fig. 47.

This tool fills a want or a neat and convenient Scribing Block that can quickly be adjusted to the finest work. It contains the patented fine adjustment similiar to that on our surface gauge, the needle being set in any position and held perfectly rigid by the side thumb nut while using the fine adjustment, by means of the eccentric thumb washer on top. It has a V-slot in the bottom of base by which work can be quickly and accurately trued to a boring bar. It also makes a useful and convenient scratch and depth gauge. Price, each $1.50. (Code word—*Unity.*)

CARR'S PATENT
Key Seat Rule Clamp.

Fig. 48.

These Clamps can be adjusted to *any* two scales or straight edges of *unequal* thickness, width or length, making a most complete and convenient Key Seat Rule or Box Square. Price, 75c. per set.
Code word, *Unique.*

Carr's Improved Centre Square.

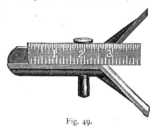

Fig. 49.

PAT. APPLIED FOR.

For many years machinists and pattern makers have needed a *Centre Square* of simple construction, which could be used with *any* straight edge or scale. This square which we now offer is simple, complete and accurate, and a straight edge or scale of *any thickness* from ½ to 1⅛ inches in width can be adjusted to it.
Price, 75 cents each. Code word, *Urge.*

MANUFACTURED BY

The Hoggson & Pettis Mfg. Co.,
NEW HAVEN, CONN., U. S. A.

ESTABLISHED 1849.

HOOKER, THOMAS, Syracuse, NY

Associated with the SYRACUSE TWIST DRILL CO., probably as the owner, Hooker received a patent for a micrometer comparator on April 29, 1884, which appears to be the first design of a comparator introduced by the SYRACUSE TWIST DRILL CO. in 1885 and covered by the Jacob Hurst patent of November 30, 1886. Hooker was also the assignee of the John E. Sweet micrometer patent of March 10, 1885. (See SYRACUSE TWIST DRILL CO. entry in *Makers of American Machinist's Tools*, first volume, for details.) Hooker remained active for many years, receiving a patent for a micrometer gage on August 8, 1916.

HUBBARD & CURTISS MFG. CO., Middletown, CT

Incorporated on April 22, 1872, as a reorganization of the Hubbard Hardware Co., maker of boxwood and ivory rules. C.C. Hubbard was president and F.W. Hubbard, secretary. The company appears to have absorbed the WARWICK TOOL CO., Middletown, CT, at that time, or at least most of their products. Its 1872 catalog offered surface gages, hack saws, and hand vises previously made by Warwick.

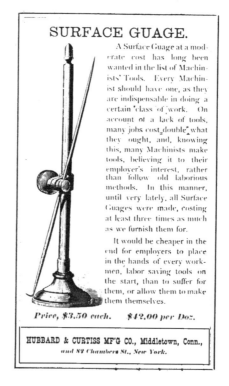

1872

HUNTOON & SIMONDS, Lowell, MA

Makers, in 1871, of a tap wrench patented February 7, 1871, by George Huntoon and Edwin Simonds. The tool is marked:

H. & S.
LOWELL, MASS.
PAT APD FOR

See also SIMONDS, Lowell, Mass.

HYNES, F.R., Camden, NY

Maker of a small surface gage with $2\frac{1}{8}$" round base and 5" post. Date of operation is unknown, but the tool appears to be from the 1880s or 1890s.

INTERNATIONAL TOOL CO., San Leandro, CA

Maker of inside micrometers in the late 1930s. These well made micrometers were furnished in a distinctive bakelite case. The same tools were offered by the PACIFIC TOOL & SUPPLY CO. about 1939.

JENNESS, JOHN S., Bangor, ME

Maker of machinist's calipers patented by Joseph W. Strange of Bangor, ME, September 11, 1860.

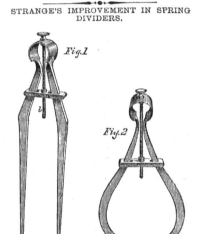

Scientific American 1860

JILSON, CLARK, Worcester, MA

See SOUTHWICK & HASTINGS in this volume and J.& H. in *Makers of American Machinist's Tools*, first volume.

J.W.E.B.

Mark found on rules made by JOHN WYKE, East Boston, MA

KENNELLY & CAIN (CRESCENT TOOL WORKS), Bridgeport, CT

The below copy of the instruction sheet packed with the Kennelly patent protractor allows us to better understand the uses of this tool. See *Makers of American Machinist's Tools*, first volume, for other information.

THE ATTENTION OF
MACHINISTS, TOOL-MAKERS AND MECHANICS
—IS CALLED TO—
KENNELLY'S PATENT PROTRACTOR.

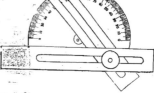

Does Not Get Out of Repair !

Is Needed by Every Mechanic !

No Tool Chest Complete Without It !

CHEAP !
ACCURATE !
SIMPLE !

The above cuts represent an improved Protractor designed especially for Machinists, Toolmakers and Patternmakers.

An experience of a quarter of a century in the machine shops of the country, convinces us that nine out of every ten journeymen have a plain bevel square of some kind. It is generally the silent evidence of their skill as an apprentice, silent, for though it may be set at any angle, yet it will not tell just how many degrees that angle is. It is here that "Kennelly's Patent Protractor" comes to his aid. That it has a wider range in determining angles, and is more practical than any other protractor, read the following :

FROM THE "AMERICAN MACHINIST," JAN. 31, 1889.

A NEW BEVEL PROTRACTOR. "The accompanying cuts show a tool, by which any machinist, or other mechanic possessing a bevel square, can also readily have a bevel protractor.

The cut to the left shows the tool by itself, while the one to the right shows its application to a bevel protractor. It consists simply of a plate graduated to degrees, and having at the center a stud which fits a hole in the plate and turns freely therein. The end of this stud, which projects on the graduated side of the plate, is cut away just to the center, as shown, and by applying this flat side of the stud to the blade of the bevel square, as shown, and then turning the blade to the proper graduation, the stud adjusts itself to the blade, which is thus set to the desired angle.

It will readily be seen that this tool, accurately constructed, and used in connection with a good bevel square, will be very convenient, and also capable of securing very accurate results, as well as making a tool, now used for one purpose only, useful for an additional purpose. The tool is made by the Crescent Tool Works, Bridgeport, Conn."

A Pennsylvania Machinist writes :—" I have found a new use for your protractor, that is, I can by its use, turn a taper piece in a lathe and have it come the right taper the first chip over. You know of course that turning a taper, is to a great extent, a cut and try affair. If you don't get the correct taper the first chip, you are obliged to set your tailstock over and try again, and so on until you have the correct taper. Often too, the machinist works at some distance from the machine to which the taper piece is to be fitted, and when the time lost in going to and from is added to the cut and try method of turning, it makes a very costly way of working. Now let me tell you how I do it. In the first place, I use a bevel square that has a blade such as is made by Starrett. I had a lathe enter to turn, and by taking out the old one and setting my bevel and getting the taper of it, and then applying the bevel to my protractor, I found how many degrees that taper was. Then by setting the taper attachment on the lathe over the number of degrees called for, I made a perfect fit the first chip. I don't know of any other protractor made that will do this, and I can say with truth that any owner of a machine shop having any taper work to do, will save the cost of the protractor ten times over in one week, even should he supply all the men in his shop with a protractor each at his own expense."

TESTIMONY OF BRIDGEPORT ARTISANS.

The undersigned having used and carefully tested Kennelly's Patent Protractor, have no hesitation in saying that for simplicity, cheapness and accuracy, it has no equal. We cheerfully recommend it to Machinists, Toolmakers, Patternmakers and others, as admirably answering the wants of all tradesmen requiring a tool of this kind.

SIGNED:

O. E. Buckminster, Master Mechanic,	Wheeler & Wilson.	Wm. DeLaney, Toolmaker,	Union Metallic Cartridge Co.
Geo. M. Eames, Asst. Manager, Machine Department,	"	Byron E. Boydon,	"
H. H. Brautigam, Toolmaker,	"	F. E. Bradley, Toolmaker Cap Department,	"
W. W. Eames, Toolmaker,	"	M. D. Holden,	"
N. S. Warner, Manager Machine Department,	"	J. M. Parkhurst	"
E. W. Judge, Toolmaker,		W. W. Ingham, Manager, "	"
George H. Woods, Foreman, Toolroom,	Union Metallic Cartridge Co.	James Cary, Toolmaker,	"
Thos. Smith, Asst. Foreman	"	John McBride, Patternmaker, Cap	
Joel Griffin, Toolmaker,	"	Edwin W. Mathewson, Toolmaker,	Yost Typewriting Co.
Eugene E. Norton, Toolmaker,	"	Thomas Kerr,	

Do not therefore borrow your shopmates tools any longer when **Kennelly's Patent Protractor** can be had for One Dollar by mail. Do not pay six times the price for a tool that is not as good for general work. It is made of steel, nicely finished and accurately graduated, and put up in a nice paper box, which will keep it from contact with other tools.

To avoid delay, address all orders and communications to

MONTGOMERY & CO., Sole Agents,
105 FULTON STREET,
NEW YORK.

Ask your nearest Hardware Dealer for it.

MANUFACTURED BY
CRESCENT TOOL WORKS,
BRIDGEPORT, CONN.
P. O. BOX 1741.

KEYSTONE REAMER & TOOL CO., Millersburg, PA

A 2 - 3" micrometer marked as above has been reported. It has features found only on micrometers made by the MASSACHUSETTS TOOL CO. or later, the GOODELL-PRATT CO.

KIMBALL & TALBOT, Worcester, MA

The below advertisement from the 1865 Worcester City Directory shows the complete line of calipers made by this company. See entry in *Makers of American Machinist's Tools,* first volume, for more details.

KLINE & HELLER, Allentown, PA

Makers of "Allen" trammel points during an unknown period. The well made points have an ingenious cam lock feature for securing the points to the bar.

LaFRANCIS, ANTHONY, Newark, NJ

Maker of a combination drill gage and stand for which he was granted a patent January 30, 1906. The tool was offered by the JAMES L. TAYLOR MFG. CO., Bloomfield, NJ, in 1905 and later by LaFrancis.

LARSON, NILS E., Chicago, IL

Maker, in 1916, of: "an adjustable male gage which has the body made of steel and adjustment of the measuring points provided by means of a wedge actuated by a knurled nut."

LE COUNT, C.W., South Norwalk, CT,

later

LE COUNT, W.G., South Norwalk, CT

Started in 1867 by Charles W. LeCount (1828-1893) to make lathe dogs, machinist's clamps, jack screws, and mandrels. Charles' son, William G. LeCount, took over the business in 1893 and continued the same products. The firm operated into the 1950s.

Machinery 1900

LeCount's Cast Iron
JACK SCREW
A strong, well made tool for use as blocking on Planers, Drill Presses, Milling Machines, etc. Made in 6 sizes.
Send for catalogue of machinists' tools.
WM. G. LECOUNT,
Post Office Box 450,   So. Norwalk, Conn.

LEOMINSTER NOVELTY & SPIRIT LEVEL WORKS, Leominster, MA

Founded by M.F. Morse in 1884 as the Leominster Novelty Works to make horn products. In 1888, he bought out the FITCHBURG LEVEL CO., moved production of the levels to Leominster, and changed the name of his company to include the spirit level product.

Frasse & Co. 1889

With double plumb and side sights	6 inches,	per dozen,	$9 00.
" " "	8 "	"	12 00.
" " "	12 "	"	15 00.
" " "	18 "	"	24 00.
" " "	24 "	"	30 00.

LINDQUIST ENGINEERING WORKS, Portland, CT

Introduced, in 1919, an unique micrometer which read directly in .0001". This was done by using two screws with opposing pitches, one of 40 tpi (.025" per turn) and the other of 50 tpi (.020" per turn). The effect was that when the thimble was turned one rotation the spindle moved only .005". The thimble was graduated to 50 divisions with the result that one division equaled .0001". Total range was ¼".

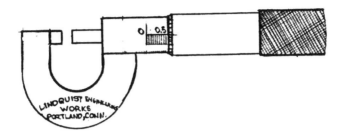

LINNAHAM, PETER, Beverly, MA

Maker, circa 1910-1912, of indicating calipers for which he was granted a patent on May 23, 1911. The unique feature is a removable leg which allows the other leg, with the indicator mechanism, to be used as a test indicator. The only known example is marked only PAT. MAY 23, 1911, and is missing the removable leg.

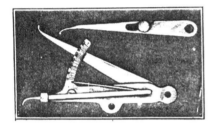

FIG. 1. An Indicating Caliper

American Machinist 1910

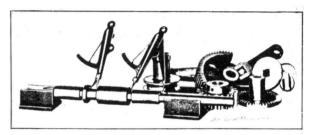

FIG. 2. Measuring between Shoulders

LONG, CHARLES B., Worcester, MA

Charles B. Long (?-1892) was the inventor of a gear-calculating rule, patented April 25, 1865. He advertised the rules under his own name in 1864 and 1865, at a price of $5.00, and under the name of the WORCESTER RULE CO. in 1866 for $4.50. By 1868 the STANLEY RULE & LEVEL CO. was offering the rules at a retail price of $4.00, but it is doubtful that they were making them. The Long rule was listed in the Stanley catalog of January 1, 1870, but it may have been in the same catagory as the complete line of DARLING, BROWN & SHARPE machinist's tools which was also listed. By 1872 the Long rule had disappeared from Stanley catalogs.

NOTE: Since the use of gear-calculating rules is an art long since lost, a short explanation might be in order. Machine gears, from early times until the last quarter of the 19th century, were nearly all cast rather than cut. Gear-caluclating rules were used by pattern makers to lay out the wooden patterns used to produce the castings, not to measure finished gears.

Long was also granted a patent for a rotating metal tube for the protection of spirit level vials on November 8, 1887. The feature, which became very common on machinists' levels, was first used by R.J. SANFORD, Worcester, MA, on his iron levels.

(See illustration next page.)

LX   See MANUFACTURER'S BELT HOOK CO.

PATENT IMPROVED GEAR, OR COG WHEEL, CALCULATING RULES,

FOR THE USE OF

MACHINISTS, ENGINEERS, MILLWRIGHTS, INVENTORS, ETC.

Two Feet, Two Fold, Boxwood, Brass Bound, Graduated the *entire length* into 5ths, 6ths, 7ths, 8ths, 10ths, 12ths, 14ths and 16ths of inches, with Tables for the easy calculation of all Gear Cutting and Cog Wheel Work.

MacDONALD, JOHN A., Lowell, MA

Maker, circa 1887-1888, of a bar micrometer for which he was granted a patent on June 19, 1888.

MANUFACTURER'S BELT HOOK CO., Chicago, IL

Maker of a variety of leather machine belt tools and accessories beginning about 1906. In 1920, they offered the speed indicator shown below. Specimens examined are marked only with LX cast into the handle.

LX BALL BEARING DOUBLE SPEED INDICATORS

Fig. 35222A

Smaller dial, automatically and accurately registers the number of revolutions either right or left. Fitted with special bearings, nickel-plated.
Price............................each 1.50

MARVIN & CASLER CO., Canastota, NY

Maker, in 1914, of a "Rotary Center Indicator" shown below. The tool appears to be the prototype for the edge finder later made by the L.S. STARRETT CO.

(See next page for illustration.)

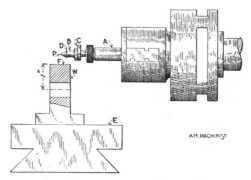

Fig. 2. Details of Rotary Center Indicator

MASSACHUSETTS TOOL CO., Greenfield, MA

The following ads and product introductions expand the information contained in the company's entry in *Makers of American Machinist's Tools,* first volume.

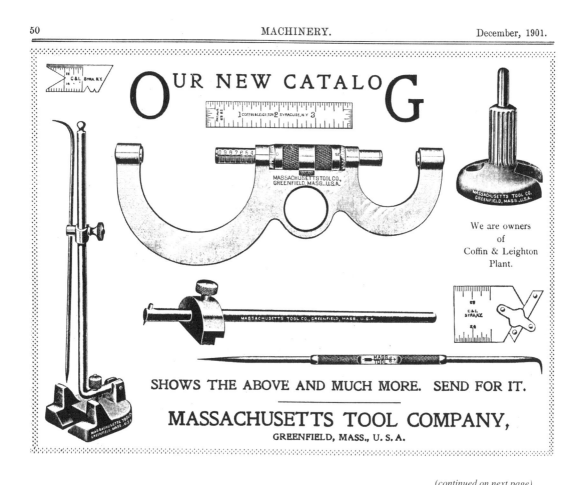

(continued on next page)

MASSACHUSETTS TOOL CO.

NEW MACHINISTS' TOOLS.

The illustrations, Figs. 8 and 9, show two tools placed on the market by the Massachusetts Tool Co., Greenfield, Mass. The first of these is a 6-inch micrometer caliper designed for measuring from zero to 6 inches by half thousandths. The sliding micrometer head travels on a cylindrical barrel through which a hole is accurately bored to accommodate three plugs, one, two and three inches long, as in the engraving.

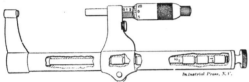

Fig. 8. Six-inch Micrometer.

These plugs serve to locate the traveling head at fixed distances one inch apart. The micrometer screw itself has a travel of one inch, like any standard micrometer. A locknut is used to hold the screw in any desired position. A thumb screw at the end of the barrel bears against the end plug and zero marks are provided to bring the screw against the plug with the same degree of pressure at each setting. When the head is clamped by means of the locking nut, it is as rigid as though it were solid with the barrel, and the faces of the measuring points are thus always parallel.

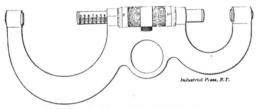

Fig. 9. Combined One and Two-inch Micrometer.

A combined one- and two-inch micrometer is shown in Fig. 9. One side records measurements up to one inch, and the other side up to two inches. A single knurled sleeve or nut serves to move the double-ended measuring piece one way or the other as desired, this piece having a travel of one inch. The spindle is non-rotating, so that the faces of the screw and anvil are always parallel. A locking device holds the screw in any position. This tool is convenient for use both in measuring and as a gage, since it can be conveniently held by the finger ring appearing at the back. A modification of the 6-inch micrometer is made in the form of a 6-inch micrometer surface gage which operates on the same principle.

Machinery January 1902

MACHINISTS' TOOLS.

The illustrations, Figs. 1 and 2, are of new machinists' tools that have been placed on the market by the Massachusetts Tool Co., Greenfield, Mass. Fig. 1 shows a depth gage with rule which is somewhat like the regular type of depth gage with narrow tempered steel rule, but has additional features. By turning the knurled nut on the back the rule can be loosened and swiveled so that it will be parallel with the base, making it convenient to carry in one's pocket or to place in a corner of a tool chest. The side of the head is graduated into 30, 45 and 60-degree angles and the rule can be turned and clamped so as to conform to any of these lines, making a convenient bevel square. It may also be used as a small T-square.

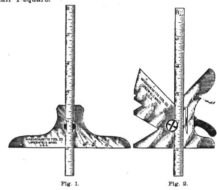

Fig. 1.   Fig. 2.

Fig. 2 shows a steel center square with rule, which is a light but very rigid tool. It may be used as a center square, center gage, depth gage or T-square. The rule is easily removed from the head. The head itself lies perfectly flat on one side of the work. The tool is small and light enough to be easily carried in one's pocket. The aim in producing these two tools has been to dispense with needless weight and space.

Machinery July 1902

McGRATH, ST. PAUL CO., St. Paul, MN

Founded in 1946 by John B. McGrath and Edward J. McGrath. The first city directory listing includes an ad listing a variety of machinist's tools including: steel rules, vernier calipers, gear tooth vernier calipers, vernier depth gauges, vernier height gauges, pocket calipers, spring calipers, micrometer calipers, squares, telescoping gauges, wire gauges, etc.

The list is so long as to raise the possiblity that they were dealer's rather than manufacturers. However, the few McGrath tools examined are clearly not products of any other known company and exhibit features unique to McGrath.

By 1951, the firm was listed as a manufacturer of tools and electric household appliances and had disappeared by 1953.

McKNIGHT, G.L., Worcester, MA

Inventor and maker of finely adjustable calipers patented April 30, 1867. McKnight received a second caliper patent May 28, 1867, but the latter patent date is not marked on existing specimens of the caliper.

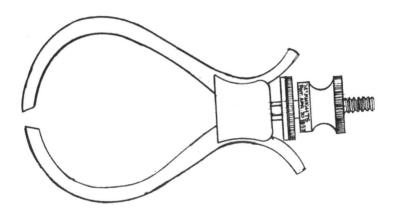

MEIGS-POWELL CO., Milwaukee, WI

In addition to the entry in *Makers of American Machinist's Tools,* first volume, we now know that the company was formed in 1920 by Arthur E. Meigs and John D. Powell. Both men had been employed by the L.S. STARRETT CO., Meigs as manager of the New York office and Powell as manager of the Chicago office.

METAL ITEMS, Racine, WI

Maker, beginning in 1949, of the "Mikro Marker" height gage shown below. The tool is well made but seldom seen.

(Illustration on next page.)

Mikro Marker Is a Micrometer Precision Height Gage

A micrometer height gage, the Mikro Marker, can be used for layout and for gaging applications. The micrometer reading range is from 0 to 1 in. by thousandths. No vernier is used. Price $12.

MICROMETER CALIPER CO., St. Louis, MO

Maker of an 1½" travel, open thread micrometer. Designed to be made at low cost, the company proudly stated: "The Cost Places It Within the Reach of All." When the entry for this company was written for *Makers of American Machinist's Tools,* first volume, it was not certain that any of the micrometers covered by the Charles E. Coe patent of May 1, 1900, were actually made. Several examples are now known, one of which is shown below. The tools are marked:

CHAS. E. COE
PAT. MAY 1, 1900
ST. LOUIS, MO

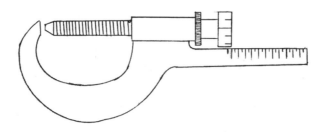

MILLIKEN MACHINE WORKS, West Newton, MA

Maker, in 1912, of machinist's V blocks and tap wrenches. By 1922, it had added surface plates and parallels.

MISCHKER, JOHN, Milwaukee, WI

Maker, in 1930, of the unusual micrometer shown below. It had an unusual quick-setting feature that allowed the square barrel to slide freely when the hook-shaped thumb piece was depressed. The barrel was also reversible, allowing the tool to be used as an inside micrometer. It appears to have been very costly to make and probably was not a commercial success. Note the odd marking of PATENTED APPLIED FOR.

It is likely that the micrometers were no longer in production by the time his patent was granted, on April 12, 1932.

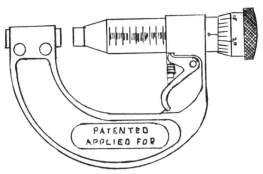

MITCHELL, WILLIAM H., Boston, MA

Inventor of a complex, worm-gear and pinion operated, fine adjusting caliper, patented July 3, 1883, and October 9, 1883. The only known example is marked with the patent dates only, so we cannot be sure that Mitchell was the maker as well as the inventor.

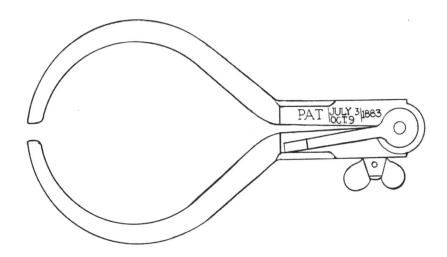

MOLDENKE, RICHARD G., New York, NY

When the entry for *Makers of American Machinist's Tools,* first volume, was written, there was doubt that the odd looking surface gage introduced by Moldenke in 1889 was actually made. We now know that it was. One of the tools has surfaced and is exactly as shown in the illustration in *Makers of American Machinist's Tools,* first volume. It is marked R.M. and PAT APPLD FOR. An 1889 advertisement from R.G.G. MOLDENKE, manufacturer of fine tools, has also been located. The selling price was $3.00 and, strangely, he refers to the tool as a "New Combination Square for Mechanics."

MONROE BROTHERS, Fitchburg, MA

Maker, about 1870, of locking calipers as shown below.

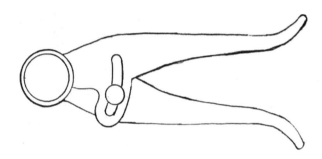

MONTGOMERY & CO., New York, NY

A large New York hardware dealer, the firm also offered a variety of machinists' tools under its own name. Others were offered under their FULTON trade name. See entry in *Makers of American Machinist's Tools, first volume* for more details.

The below excerpts from their 1901 catalog show thread gages, screw gages and drill gages marked MONTGOMERY & CO. and speed indicators with the FULTON trade name. The speed indicators were made under William Lang patents of November 18, 1882, and October 9, 1894, and had been previously offered by CHURCH & SLEIGHT, New York, NY. The gaging center punch was made under the John E. Goodard patent of October 4, 1898, and is marked only with that date.

(continued on next page)

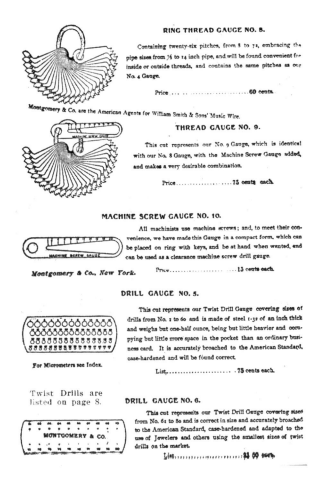

MORSE TWIST DRILL & MACHINE CO., New Bedford, MA

Founded in 1864 by Stephan A. Morse (1828-1898). Maker of the first commercially available twist drills, patented by Morse in 1863. A very interesting twist drill gage, shown below, was found at the factory when it was closed and the contents auctioned a few years ago. Dated November, 1865, and marked S.A. MORSE, a case could be made that this was the prototype twist drill gage. See *Makers of American Machinist's Tools*, first volume, for details on later Morse drill gages.

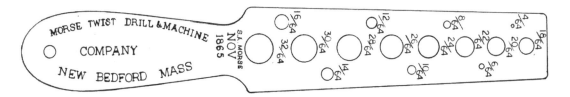

MORSE, WILLIAM A., Boston, MA

Inventor and maker of self-registering calipers patented November 8, 1863. The calipers were offered for sale in 1864 and 1865.

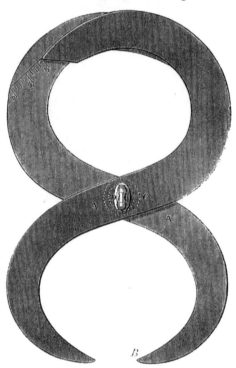

Scientific American 1863

MORSE'S SELF-REGISTERING CALIPERS.

Perhaps the tool most universally employed by machinists is a pair of calipers; from taking the size of a drill or a rod, up to turning a shaft, they are in constant requisition, and are quite indispensable. The instrument herewith illustrated is of the class known as registering calipers, and by a very simple arrangement of a scale on each side of one pair of the legs, A, the distances of the points, B, are accurately measured. This is a very convenient form of self-registering calipers, as the workman can see, by a glance at the scale, the size required, without being obliged to carry a rule in his pocket.

MOSS-OCHS CO., Cleveland, OH

A folding radius gage, 1/8" to 1" by eighths, marked as above, has been reported. It's likely that the name is a misreading of a poorly marked tool from radius gage maker F.G. MARBACH CO.

MOSSBERG, FRANK, Pawtucket, RI

Inventor and maker of a micrometer gage, patented January 25, 1887. Mossberg was a maker of drill-grinding machines in 1887, when he began offering the gage. In 1889, he founded the Mossberg Mfg. Co. in Attleboro, MA.

(See next page for illustration.)

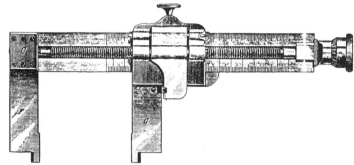

Novelties.—Fig. 6.—A New Micrometer Gauge, Made by Frank Mossberg.—Top View.

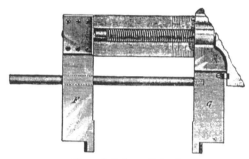

Fig. 7.—Depth Gauge Attachment.

MOTOR ENGINEERING CO., Cleveland, OH

Maker, in 1914, of a locator (we would now call it a wiggler). The same tool is still being made by the L.S. STARRETT CO. as their #828 wiggler or center finder.

American Machinist 1914

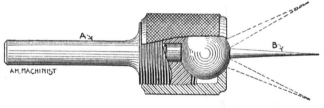

Fig. 1. The Locator

N.D.T. CO., Boston, MA

Maker of a combined drafting and measuring tool, patented March 28, 1899, by William C. Day of Rumford Falls, ME. The tool consists of dividers with a third leg that stays centered between the two outer legs.

NASH ENGINEERING CO., Brooklyn, NY

Operated by Louis H. Nash, the company was the maker, in 1910, of a micrometer plug gage patented by Nash on March 29, 1910, and "designed to measure small holes or slots as accurately as the ordinary outside micrometer will measure external diameters." The barrel was graduated with 50 divisions, each of .0002", and the tool had a total range of $\frac{1}{16}$".

American Machinist 1910

A Micrometer Plug Gage

NATIONAL BUREAU OF STANDARDS, Washington, DC

The function of the National Bureau of Standards is to establish and police various standards used by business and industry. On at least one occasion, however, the Bureau entered into the manufacture of measuring devices.

During World War I, probably due to fear of American industry being cut off from Johannson gage blocks, then made only in Sweden, the Bureau produced gage blocks equivalent to the Johannson blocks.

William E. Hoke, who had developed a process and machinery for the manufacture of gage blocks, was hired to oversee the installation of machines, and the later manufacture of gage blocks at the Bureau. A number of sets were made and distributed to industry by 1919.

After the war Hoke, with his machinery and process, moved to the PRATT & WHITNEY CO. which continued making HOKE blocks for many years.

Blocks made at the Bureau are identified by their circular cross section. Blocks made by PRATT & WHITNEY have a square cross section.

NEW HAVEN SPECIAL TOOL CO., New Haven, CT

Maker, about 1890, of a scribing block designed by James Carr. The only known example is marked with the company name and PAT APPLD FOR and has a fine adjustment mechanism slightly different than that used in later production. By the time the patent was issued on January 26, 1892, the tools were being made by the HOGGSON & PETTIS MFG. CO., New Haven, CT. The company may also have made the early production Carr patent surface gages, which are marked only PAT APPL FOR and differ in many details from those marked with the HOGGSON & PETTIS MFG. CO. name.

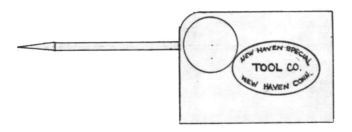

NICHOLSON FILE CO., Providence, RI

Founded by William T. Nicholson (1834-1893) in 1864. In addition to a wide variety of files, the company made machinist's file holders and file handles under Nicholson's patents of June 12, 1877, January 1, 1878, and June 4, 1878. See W.T. NICHOLSON for earlier products.

NICHOLSON, W.T., Providence, RI

Operated by William T. Nicholson (1834-1893), who began making milling machines, vises, and iron levels in 1859. The levels were made under his patent of May 1, 1860. In 1864 he sold this business and formed the NICHOLSON FILE CO. Nicholson sold or licenced the level line to the STANLEY RULE & LEVEL CO. which continued making them until 1892.

NORRIS, SAMUEL, Springfield, MA

Maker of calipers and dividers patented July 21, 1863, by C.A. Fairfield. The design, shown below, was also used by J.STEVENS & CO. after the patent expired.

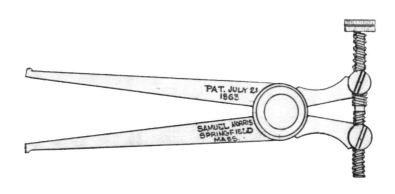

OSBERG & JOHANSON, Dorchester, MA

Makers, in 1919, of the "True" indicator. The device was touted as: "has no delicate springs, levers or other small parts to get out of order." The tool was, in fact, a wiggler rather than an indicator, although it did include a separate scale to measure the motion of the wiggler pointer. It is unclear how reliable measurements could be made in that manner.

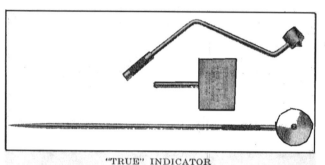

American Machinist 1919

"TRUE" INDICATOR

PACIFIC TOOL & SUPPLY CO., Los Angeles, CA

Maker, circa 1939-1941, of inside micrometers and telescoping gages. The inside micrometers were previously offered by the INTERNATIONAL TOOL CO.

PAYLER, J.W., Detroit, MI

Maker, in 1894, of file handles for which he was granted a patent on April 5, 1892. These were "devised as a result of his experience of some years as a machinist." The holder in Fig. 4 below is especially interesting as it was "designed for finishing keyways in hubs of large wheels, connecting rods, etc., one, or two workmen working together, being able to use an ordinary file to better advantage than special files usually employed for such purposes."

(See illustration on next page.)

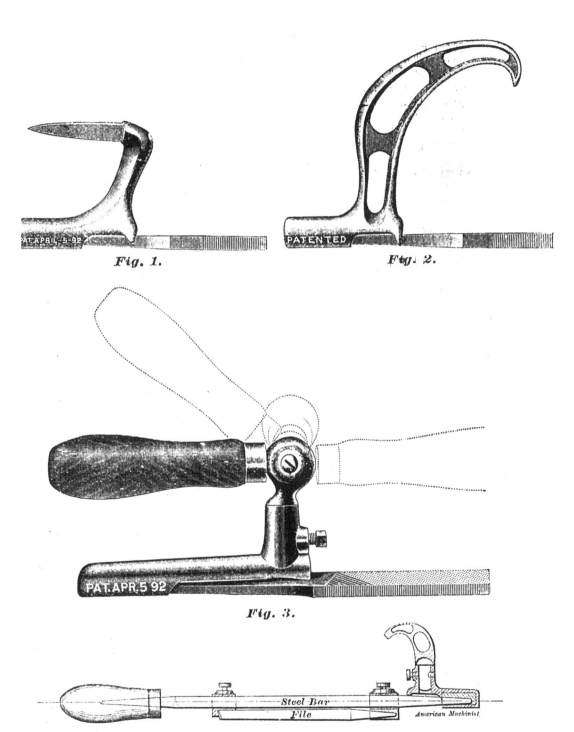

Fig. 4.
SPECIAL HANDLES AND TOOL HOLDERS.

American Machinist 1894

PITTSBURG INSTRUMENT AND MACHINE CO., Pittsburg, PA

Maker, in 1915, of combination inside-outside calipers. The calipers appear to be closely related to calipers made by the PHOENIX ROLL WORKS, Pittsburgh, PA, at an earlier date.

Lock-Joint Caliper

The illustration makes clear the construction of the type of caliper shown. While it can be applied to a wide variety of common work, it is especially adapted for

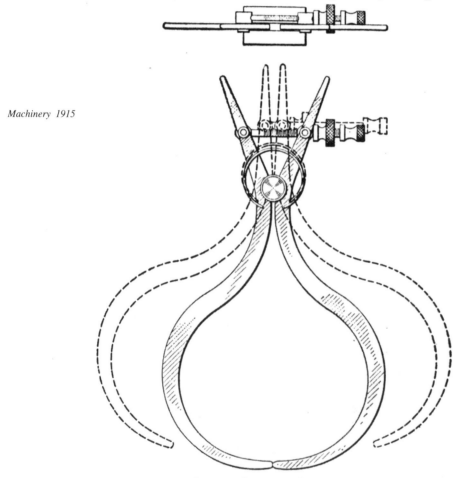

Machinery 1915

Lock-Joint Caliper

calipering the inside of chambered cavities, over flanges and the like, because it can be removed and replaced without losing the size calipered.

This form of caliper represents a recent product of the Pittsburgh Instrument & Machine Co., Pittsburgh, Penn.

PRATT & WHITNEY CO., Hartford, CT.

The following four pages from the 1893 Pratt & Whitney Co. small tool department catalog illustrates the knife edge straight edge sets and Kidd's dividers which were taken over by Pratt & Whitney from the HARTFORD TOOL CO. in 1890. The knife edge staight edge sets were eventually taken over by the BROWN & SHARPE MFG. CO. See PRATT & WHITNEY CO. and HARTFORD TOOL CO. entries in *Makers of merican Machinist's Tools,* first volume, for more company history.

SMALL TOOL DEPARTMENT. PRATT & WHITNEY CO., HARTFORD, CONN. 53

Standard "Knife Edge" Straight Edges.

THE cut represents a set consisting of a 7-inch test bar, and three straight edges, $6\frac{1}{2}$ inches, $4\frac{1}{2}$ inches and $3\frac{1}{4}$ inches long, respectively, with non-conductor.

The test bars are made of glass, cased in felt and leather, as a protection against accidents and changes of temperature caused by handling. The straight edges are made of tool steel, in shape most convenient for use, combining strength with lightness and hardened at the straight edge only.

A non-conducting handle, fitting all three sizes of straight edges, will be furnished, if desired, and is essential when these tools are used on work requiring great accuracy.

They are finished at a temperature not lower than 75 degrees or higher than 85 degrees Fah., and will be found most accurate within that limit.

Glass is adopted for the test bars, as it is less affected by changes of temperature than any other available material.

PRICE-LIST.

Set, in velvet-lined case, consisting of test bar and three straight edges; weight, 20 oz.	$12.00
Test bar, 7 inches long, in cloth-covered box	5.50
No. 1 Straight Edge, $3\frac{1}{4}$ inches long, in cloth-covered box	1.50
No. 2 Straight Edge, $4\frac{1}{2}$ inches long, in cloth-covered box	2.00
No. 3 Straight Edge, $6\frac{1}{2}$ inches long, in cloth-covered box	2.75
Non-conducting handle for straight edges; weight, $8\frac{1}{2}$ oz.	.50
Velvet-lined case for set complete; weight, $8\frac{1}{2}$ oz.	1.00
Cloth-covered boxes for test bar	.20
Cloth-covered boxes for straight edges ....each,	.15

Weight straight edges, No. 1, 1 oz.; No. 2, $1\frac{1}{2}$ oz.; No. 3, 3 oz.

Weight cloth-lined boxes, No. 1, $\frac{1}{2}$ oz.; No. 2, $\frac{3}{4}$ oz.; No. 3, $1\frac{1}{4}$ oz.

Weight glass test bar, $6\frac{1}{4}$ oz.

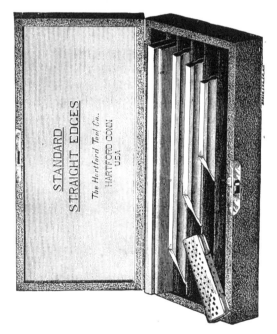

STANDARD "KNIFE EDGE" STRAIGHT EDGES.

PRATT & WHITNEY CO., HARTFORD, CONN.   55

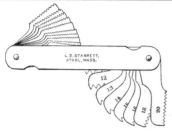

Improved Screw Pitch Gauge.

THIS gauge has twenty pitches, viz.: 9, 10, 11, 12, 13, 14, 15, 16, 18, 20, 22, 24, 26, 28, 30, 32, 34, 36, 38, 40. This is the *only* gauge made that can be used inside a nut, as well as on the outside of a screw or bolt. Notwithstanding its double use and value, it is sold for the same as those of the old style having less pitches. Price, $1.00.

Center Gauge.

AND GAUGE FOR GRINDING AND SETTING SCREW TOOLS.

HALF SIZE.

THE angles used in this gauge are 60 degrees. The four divisions upon the gauge of 14, 20, 24 and 32 parts to the inch, are very useful in measuring the number of threads to the inch of taps and screws. The following parts to the inch can be determined by them, viz.: 2, 3, 4, 5, 6, 7, 8, 10, 12, 14, 16, 20, 24 and 32. Price, 50 cents.

Also on hand, Center Gauges of the Whitworth or English Standard, 55 degrees.

Inserted cutter counterbores of above style will be furnished to order, at prices to be given on receipt of specifications and all dimensions of guide, cutter-head, neck and shank.

---

54   SMALL TOOL DEPARTMENT.

Kidd's Improved Divider.

SOME OF THE SPECIAL ADVANTAGES OF THIS TOOL ARE:

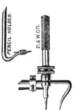

1st. It takes the place of from two to five pair of ordinary dividers.

2d. There is theoretically no limit to circles that can be drawn.

3d. It is quickly and easily adjusted to different size circles and spacing.

4th. It is made of the best tool steel, and points carefully hardened.

5th. It can be used with pencil.

6th. It is indispensable for cutting circles in soft metals for models, etc.

7th. The points are parallel and vertical while in use, making it very easy to set it to a scale.

As all workmen know, who have occasion to use dividers, it is difficult to lay out a perfect circle on steel, or other hard metal, with the ordinary divider, and only men accustomed to the work can do it, on account of the point in center punch mark jumping out as the pressure is put upon the point doing the work. Not so with this tool; the most inexperienced workman can use this, it is so rigid, the point of resistance so near the point doing the work, and as they remain parallel and vertical, little chance is given for the points to either jump out or spring away from the desired position.

Divider, with bars $1\frac{3}{4}$ and 5 inches long, in box; weight, $2\frac{1}{2}$ oz..Price.

With 5-inch bar, 7-inch circles may be drawn.

Bars, $1\frac{3}{4}$ inches long, $\frac{1}{4}$ oz........................Price,
Bars, 5 inches long, $\frac{1}{2}$ oz..........................Price,
Long bar, for 24-inch circle; weight, 1 oz............Price,
Lead pencil holder, short; weight, $\frac{1}{2}$ oz............Price.
Lead pencil holder, for 24-inch circle; weight, 1 oz....Price,

PRICES.—Adjusting nut, 10 cents; binding screw, 8 cents; justing screw, with yoke, nut and binding screw, 30 cents. binding nut, 20 cents.

Extra length bars made to order; weight of box, $\frac{3}{4}$ oz.

RAND & HAKEWESSELL, Hartford, CT

Makers of machinist's protractor patented April 8, 1890, by Reinhold Hakewessell and Fred K. Rand. Examples are marked PAT APPLD FOR. Hakewessell was co-founder of the Standard Mfg. Co., Hartford, CT, in 1888 and co-inventor of the famous Acme screw machine in 1892. The protractor business must have been a short-lived sideline, as examples are extremely rare. This firm was erroneously listed as LANE & HAKEWESSELL in *Makers of American Machinist's Tools*, first volume.

(See illustration on next page.)

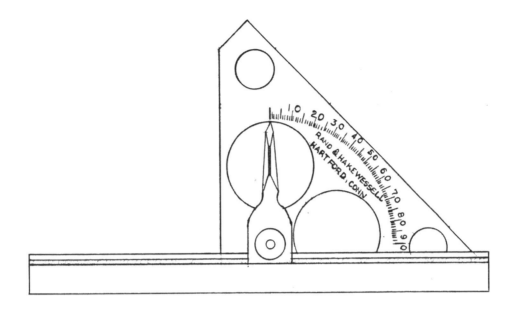

RANDALL & STICKNEY, Waltham, MA

A partnership of Frank E. Randall (1881-1941) and his brother-in-law Francis G. Stickney, formed in 1896. The new firm continued the business of making dial indicators, which had been started by Frank E. Randall in 1888.

REID, DANIEL C., Philadelphia, PA

Maker, in 1919, of Reid's Ellipsograph. He noted that: "by detaching the slotted bar it can be used as an ordinary spring divider."

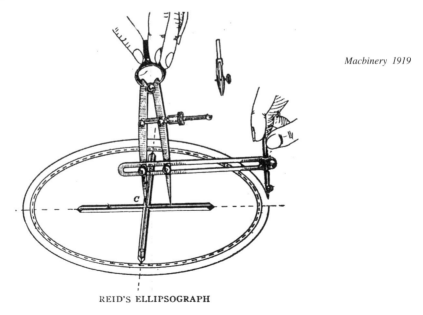

Machinery 1919

REID'S ELLIPSOGRAPH

REIFFEL & THORN, New York, NY

A partnership of Charles Reiffel and Nickolaus Thorn, makers of finely adjustable calipers under their joint patent of November 21, 1848. This patent is the earliest known for a machinist's tool and covers a tool which is among the first commercially made for American machinists. The specimen shown below is marked "PATENT APPLIED FOR," showing that production began some time before the patent date.

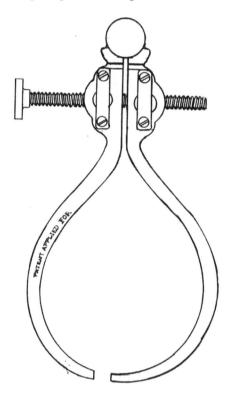

REMINGTON TOOL & MACHINE CO., Boston, MA

One of its drill grinding gages has now been examined and is obviously the same as the gages made by the HJORTH LATHE & TOOL CO. Henry Hjorth bought the plant and business of the REMINGTON TOOL & MACHINE CO. in 1912 and formed the HJORTH LATHE & TOOL CO. to continue making the same line of small lathes and machinist's tools.

In light of the above, we now know that the micrometer surface gage offered in the 1909 REMINGTON TOOL & MACHINE CO. catalog was the Hjorth patent gage later offered by the HJORTH LATHE & TOOL CO. See entries for BEST TOOL CO. and MECHANICS TOOL CO. in *Makers of American Machinist's Tools,* first volume, and HJORTH LATHE & TOOL CO. in this volume.

RENSHAW, JODEPH B., Hartford, CT

Inventor and maker of a micrometer depth gage, patented February 2, 1887. The known example is marked only with the patent date.

(See illustration on next page.)

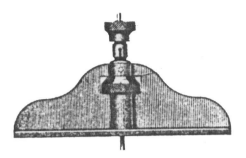

RICHARDS, FRANK H., Troy, NY

Inventor and possibly maker of the beam micrometer shown below. Richards (1850-1933), granted a patent for the micrometer September 30, 1884, was a respected machine tool designer and was granted dozens of patents for machine tools, assigning most of them to the PRATT & WHITNEY CO. Whether the tool was actually made is uncertain, but the design appears practical and Richards, with his connection to PRATT & WHITNEY CO., could easily have had some made.

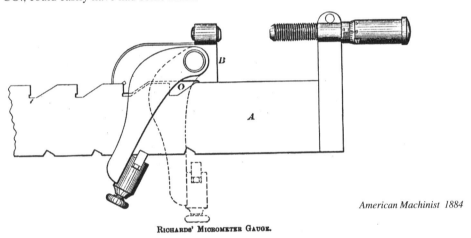

American Machinist 1884

ROBINSON CO., C.E., Orange, MA

Maker, circa 1913-1918, of a machinist's test indicator patented by Percey G. Wheeler, October 16, 1909. Also see M.B. HILL MFG. CO.

American Machinist 1916

SANDOW TOOL CO., address unknown

Maker of calipers which were copies of the L.S. STARRETT CO. "Yankee" and surface gages with a unique fine adjustment mechanism. Nearly identical surface gages marked ATLAS TOOL CO., NYC, and others completely unmarked are known. See ATLAS TOOL CO. entry for an illustration.

SANDWICH ELECTRIC CO., Sandwich, IL

Maker of a machinist's test indicator patented by William A. James on April 6, 1909. The tool is marked:

THE UNIVERSAL TESTOMETER
JAMES PAT. NO. 179,444, APR. 6, 1909
MFG'D BY THE SANDWICH ELECTRIC CO. SANDWICH, ILL. U.S.A.

The patent number as marked is transposed; the correct number is 917,444.

SANFORD, R.J., Worcester, MA

Operated by Robert J. Sanford, maker of lathes and iron machinist's levels with an adjustable protractor. The levels were furnished with a rotating metal sleeve to protect the vials, a feature patented November 8, 1887, by CHARLES B. LONG. Sanford operated from about 1880 until moving to Providence, RI, in 1891.

SAUTER, EDWARD, Hartford, CT

Inventor of a bar micrometer patented September 25, 1883. Sauter advertised the patent "For sale or let on royalty" in 1884. He further stated: "Advertiser is a good machinist, has experience in manufacturing; would engage to work for the party, or would take partner with capital, to manufacture the caliper." It would seem that he had not been successful in marketing the tool, so it is possible that very few were made.

CALIPER GAUGE.

A caliper gauge arranged to be set by means of a screw and micrometer gauge graduations, whereby both internal and external measurements may be accurately made, has recently been patented by Mr. Eduard Sauter, of Hartford, Conn. One end of the beam is bent so as to form a right angle, and the inner end of the arm is reduced in size. A sliding jaw placed upon the long arm of the beam has an arm similar in shape to the first arm and placed parallel with it. A block or yoke also slides upon the long arm of the beam, but may be made fast to it by passing a pin through it and through either one of the holes in the beam, which are one inch apart. This permits the measurement of either long or short distances. To the rear edge of the yoke is secured a sleeve which is graduated and which forms the stationary graduation of the tool. The screw rod extends through the sleeve, the yoke being tapped to receive the rod, whose front end is secured to the sliding jaw by a suitable nut and collar. On the other end of the screw rod is a revolving graduated sleeve, which, in connection with the sleeve on the yoke, constitutes the complete graduation of the tool. The sleeve on the screw rod surrounds the other sleeve and is clamped to the rear end of the screw by a nut (the construction will be readily understood from Fig. 2, which is a section through the screw rod), and is milled to facilitate turning for moving the jaw forward and backward. By the nut on the end of the rod, the rod may be adjusted while assembling the parts of the tool, or any wear occasioned by use may be taken up. The lower edge of the yoke is split and a screw inserted in order that the parts may be drawn together and clamp the screw rod, if this should be rendered necessary by wear. The jaw and sliding yoke are each provided with thumb screws by which they may be held fast to the beam. Fig. 1 is a side elevation of the gauge.

1.

SAWYER TOOL MFG. CO., Fitchburg, MA, later Ashburnham, MA

The below ads and product introductions expand on the information contained in the company's entry in *Makers of American Machinist's Tools,* first volume.

The three pages from the 1913 catalog illustrate a combination level and plumb bob, automatic two-stroke center punch, and Cook's turret head tap and drill holder. These three tools are not shown in any previous catalog and all are very uncommon. In fact, the combination plumb-bob and level has not been observed; perhaps they are all hiding in plumb-bob collections.

Also of interest is the following excerpt from *Machinery* magazine of September 1917. The excerpt is from an article on pantograph graduating machines made by John Hope & Sons, Providence RI.

> "An idea of the rate of production will be gathered from the fact that in the plant of the Sawyer Tool Mfg. Co., Ashburnham, Mass., girls who operate these machines each mark 200 feet of scales per day... When graduating steel scales, 48 six-inch scales may be operated on simultaneously. There are four work holders, each with a capacity of holding six feet of work, i.e., twelve six-inch scales."

Unfortunately, the article does not tell us the number of "girls" working, but the pictures show at least four machines. This implies 800 twelve-inch scales or 1600 six-inch scales per day; a much higher production than surviving specimens would lead us to believe. The survival rate of machinist's tools may be a great deal less than we thought. *(Illustrations are continued on the next three pages.)*

A NEW SURFACE GAGE.

The Sawyer Tool Mfg. Co., of Fitchburg, Mass., have brought out a new surface gage provided with micrometer adjustment. The sides of the base are ground parallel and true and upon the upper surface are two bosses for use in planer

Improved Surface Gage.

bed slots or similar positions. The base is slotted so that the spindle can be revolved through three-quarters of a circle, thus giving a wide range of adjustment. The clamp upon the spindle holds a block which can be swung either up or down to any desired angle and clamped solidly in position. A knurled thumb screw, on the under side of this block, holds the needle or scriber in position; while a knurled head, on the upper side, gives the needle an up-and-down parallel motion. This motion will be appreciated as a very desirable feature of the gage. The knurled head in front clamps the needle rigidly when the adjustment has been made.

Machinery 1902

The Sawyer Screw Pitch Gauge.

The Sawyer Tool Mfg. Company, Fitchburg, Mass., is putting on the market an entirely new style of screw

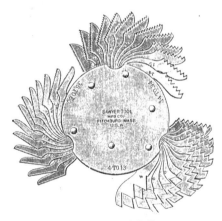

Sawyer Screw Pitch Gauge.

pitch gauge, which concentrates within a very small space and in convenient form a complete range of gauges from 4 to 60 threads to the inch. As will be noted in the illustration the gauges are in three groups and pivoted between the two disks of the case, so that all may be turned back out of the way excepting the one in use. There are 13 gauges in each group, and on the case is stamped the range of each, one group running from 4 to 13, another from 14 to 34 and the third from 36 to 60.

Iron Age 1907

SAWYER TOOL MFG. CO.

American Machinist 1897

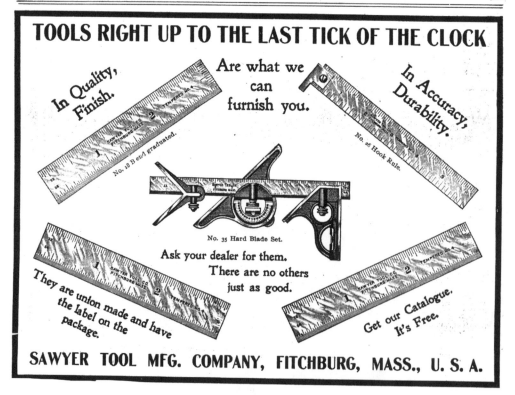

The Very Latest Thing
IN SURFACE GAUGES

Micrometer Adjustment

Just the Gauge You Want

Our New Catalogue Sent Free. Ask for it

SAWYER TOOL MFG. CO.
FITCHBURG, MASS.

July 1903

SAWYER BEVEL PROTRACTOR No. 95

An accurate, inexpensive tool for describing or obtaining any angle with the use of an ordinary straight-edge, scale or bevel. Has the additional features of positive 45° and 90° angle. This tool is adapted to all trades. Our new 64 page catalogue is ready. Sent on application.

SAWYER TOOL MFG. CO.,
82 Winter St., Fitchburg, Mass.

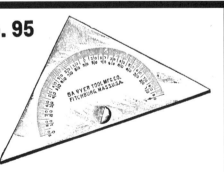

November 1905

SAWYER TOOL MFG. CO.

ILLUSTRATED PRICE LIST

July 1, 1913

SAWYER HIGH GRADE MACHINISTS' FINE TOOLS
(UNION MADE)

THE SAWYER TOOL MFG. CO., Inc.
ASHBURNHAM, MASSACHUSETTS
U. S. A.

Combination Level and Plumb Bob
No. 10

Same as No. 9 with end drilled so as to admit a chalk line being inserted, which converts the tool into a plumb bob, the combination of level and plumb bob being extremely useful.

Price, No. 10—3 1-2 in. long, $ .60

Automatic Two-Stroke Center Punch
No. 153

Makes either a light or heavy mark.

The tool is set for the stroke by pressing against any suitable object, and the trigger automatically engages for either a light or heavy stroke, depending on the distance the ram is pressed back.

After the tool is set, the point may be carefully placed on the work where the mark is desired and a light pressure of the thumb on the trigger releases the ram. The operation being without heavy pressure on the tool and without using a hammer, there is no opportunity for slipping, making it the surest, most accurate and most convenient center punch on the market.

Its advantages are appreciated where light is poor, where uniformity of marks is desired and where absolute accuracy is essential.

Points are easily replaced or removed for re-grinding.

Price, No. 153 . . . . $1.50

Extra Points, each $0.15

Cook's Turret Head Tap and Drill Holder
No. 133

Tool has six different sizes of milled square holes ranging from 1-16 inch square to 7-32 inches square.

Turret head is one inch in diameter, tool steel hardened and drawn to straw color. The head has an adjusting range of 5-32 inches and will hold drills of all sizes, from No. 50 to 1-4 inch.

Clamp Nut is of tool steel, hardened and drawn, and is provided with a wrench for use when clamping drills, round shanks, etc.

The holes being diametrically opposite, a center for trueing taps or drills is always available.

The tool makes a valuable screw driver handle, using round drill rod steel of different sizes and lengths for bits. Square one end like a Tap or Reamer, mill or file the other end to shape desired; temper and draw to proper color for greatest strength.

Can be used as a holder for tapping, reaming or drilling. It will hold the largest assortment of different sizes of taps, reamers, drills, screw driver points, etc., of any tool now on the market.

Price No. 133, complete $1.80
Extra Wrench . . . . 10c Net

SCHAEFFER & BUDENBERG CO., Brooklyn, NY

Maker of the "American" speed indicator. The tool is a true tachometer, reading speed directly from 20 to 8000 RPM in four ranges, selectable by a mechanical dial. Date of manufacture is uncertain, but appears to be circa 1900-1920. The firm also made or imported multi-range tachometers in the 1890's and later. The example shown below is fairly common and is usually found cased with a number of accessories. Note the $60.00 price tag; a lot of money in 1892.

Chas. A. Turner 1892

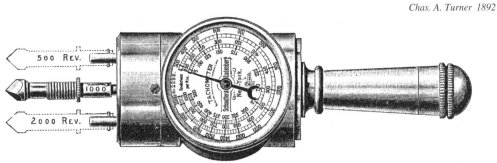

Fig. 745.

The Portable Tachometer illustrated herewith, is similar in construction to that of our Tachometer for permanent attachment. By applying this instrument by hand to the center of rotating shafts, it will instantly and correctly indicate the number of revolutions of the shaft per minute.

On unscrewing the end of handle will be found a set screw, by means of which any lost motion may be taken up when it becomes necessary. The dimensions of the Portable Tachometer are:

Diameter of dial... 2½ inches.
Depth of case... 3 "
Entire length, including handle and detachable point...................11 "
Total weight, including Morocco case.. 5 lbs.

The instrument is of excellent workmanship, plated and finished. It is contained in an elegant Morocco case, for easy transportation. We confidently recommend it for its adaptability in obtaining accurate indications of the variations in speed of fast running machinery.

The Portable Tachometer with 3 scales permits of the reading of the actual number of indicated revolutions without division or multiplication. No exchange of gears is necessary. The point only is detachable and should be fixed upon that one of the three shafts the mark of which corresponds with the scale to be used. The scales are laid out as follows: 40 to 200 revolutions, 120 to 600 revolutions; 600 to 3000 revolutions.

Price,   -   -   -   :   -   -   - $60.00.

SCHELLENBACH-HUNT TOOL CO., Cincinnati, OH

Formed in 1908 when W.P. Hunt bought out J.W. Darling's share of the Schellenbach & Darling Tool Co., becoming a partner with Charles Schellenbach. See *Makers of American Machinist's Tools*, first volume, for more information.

SCHRIER, CHARLES A., Holyoke, MA

Inventor of an "improved universal square" patented May 8, 1877. The tool is clearly a modification of the Ames universal square made by DARLING, BROWN & SHARPE at that time.

(See illustration on next page.)

IMPROVED UNIVERSAL SQUARE.

In the old form of universal square much annoyance was experienced when work was necessary to be scribed at places

Scientific American 1877

that came immediately under the cross-bar. This was particularly the case when small square or round pieces of work were to be centered.

The accompanying engraving represents an improvement upon this square, which consists of a curved or raised portion in front of the cross-bar, and so made that the space formed above the blade is sufficient to admit of drawing or scribing a line along its whole length without removing the square.

SHERWOODE TOOL & INSTRUMENT CO., Westtown, PA

Maker, in 1910, of the Knight beam compass. The compass was made with a graduated steel beam and was furnished with pencil and pen attachments for draftsman's use and with steel scribers for shop use.

SIMONDS, EDWIN, Lowell, MA

Maker, about 1872, of a tap wrench patented February 7, 1871, by George Huntoon and Edwin Simonds. The same wrench was made in 1871 by HUNTOON & SIMONDS, Lowell, MA.

SIMPLIFIED RULE CO., Philadelphia, PA

Maker of machinist's rules patented by Louis Robidoux on July 28, 1931. Through a unique method the company managed to label each 64th over a six inch scale. Despite the company's name, the result was a complex and difficult to use rule which was probably not very popular.

SMITH, EDWARD C., Brooklyn, NY

Maker of machinist's calipers, finely adjusted by a worm gear. Smith received his patent April 4, 1882. The specimen examined, shown below, is marked only with the patent date.

(See next page for illustration.)

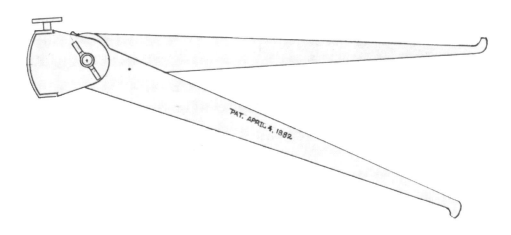

SMITH, J. CECERI, London, England

Although I did not intend to include English tool makers in this book, an exception has been made for J. Ceceri Smith. His direct reading micrometer, patented in the U.S. April 11, 1893, was one of the first to be successful, and several examples have been found in the U.S. He seems to have had enough success in the U.S. market to cause some concern at the BROWN & SHARPE MFG. CO. Brown & Sharpe bought all U.S. rights to the micrometer about 1914 when it was ready to introduce its Spaulding design direct reading micrometer. It never produced the Smith micrometer, most likely buying the rights just to eliminate a competitor to their new product.

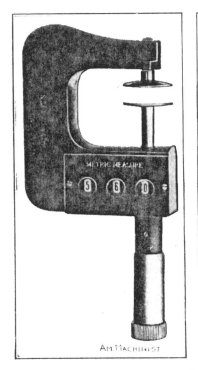  

American Machinist 1912

THREE TYPES OF CALIPER GAGES

SOMERS, J.L. & G.A., Brooklyn, NY

When the entry for *Makers of American Machinist's Tools,* first volume, was written, no illustration of their odd combination caliper and plumb was available. The tool, looking like some strange sea creature, is shown below.

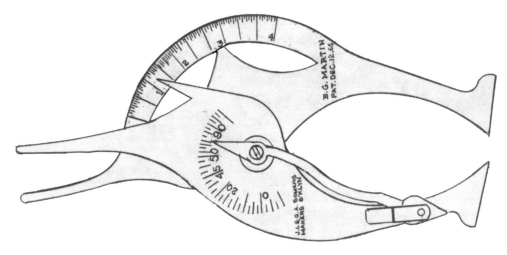

SOUTHWICK & HASTINGS, Worcester, MA

Makers, in 1868, of machinist's combination dividers and inside/outside calipers patented by Clark Jilson, August 16, 1864. The calipers were expensive, selling, in 1868, for $1.10 for the 3" size and 65c for the 1½" size. The Jilson patent calipers, shown below, are also found marked J.& H., WORCESTER, MASS. (see entry in *Makers of American Machinist's Tools,* first volume) It is believed that J.& H. was the first maker and SOUTHWICK & HASTINGS the second.

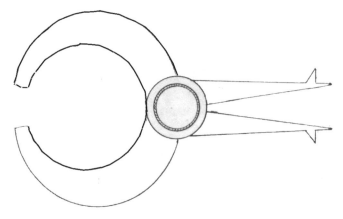

SPENCER, WILLIAM D., Middletown, CT

Inventor and maker of depth gages patented March 13, 1888. The specimen illustrated below is fully nickel plated and marked only "PAT APL'D FOR". The rule is calibrated in 32nds on one side and 64ths on the other. *(See next page for illustration.)*

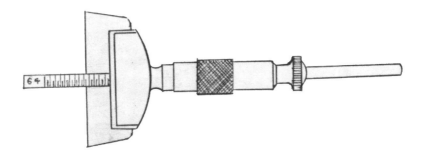

SPRINGFIELD LEVEL & TOOL CO., Springfield, MA

The following announcement appeared in the *American Machinist* magazine, November 1, 1894.

"The Springfield Level & Tool Co. have recently commenced the manufacture of a line of Carpenters' and Machinists' Levels and will make a specialty of Solid Set Levels. The goods are of new design and, it is claimed, will excel in accuracy. An inclinometer also is in preparation and will soon be on the market. J.P. Noonan, for a number of years with Davis Level & Tool Company, is manager of the new concern."

The firm was listed in the Springfield City Directory from 1895 to 1907.

STANDARD TOOL CO., Athol, MA

The following excerpt from the 1885 John Wilkinson Co., Chicago, IL, catalog shows a previously unrecognized tool made by this company. The product introduction piece for the Bellow's June 8, 1880, patent "Rapid Transit" wrench is from an 1892 American Machinist magazine and shows that the wrench was introduced later than previously believed.

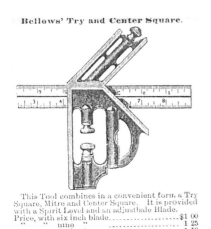

John Wilkinson Co. 1885

(Second illustration on following page.)

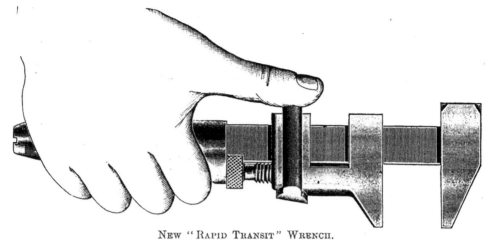

NEW "RAPID TRANSIT" WRENCH.

The bar of the wrench is made of open-hearth steel, and is case-hardened. This gives a stiff, strong bar, and one not liable to get bruised or otherwise injured.

The wrench we have seen is a well made, substantial tool, and very conveniently operated. The manufacturers are the Standard Tool Co., Athol, Mass.

STAR TOOL WORKS, New York, NY

Maker of machinists' calipers and dividers which were a copy of the L.S. STARRETT CO. "Yankee" design. Period of operation is uncertain but probably circa WWII.

STARK TOOL CO., Waltham, MA

Operated by John Stark who founded the company in 1862 to make small machinery for the watch-making industry. About 1888, he began offering the ALMORTH Universal Protractor, patented August 7, 1888, by Gustaf Almorth. Almorth assigned one-half of the patent to Stark. The protractors are very uncommon and probably were made for only a short time.

STARRETT CO., L.S., Athol, MA

The following ads expand on the information contained in the company's entry in *Maker's of American Machinist's Tools*.

The 1912 ad for the caliper with patent rule attachment is of special interest. The tool was made under the F.A. Hatch patent of April 28, 1903, and has been seen only in this ad. No tools have been observed and no one questioned has seen or heard of the tool.

(Illustrations are on the following three pages.)

NEW
Outside Spring
CALIPERS
with patent rule attachment

It is so made that zero will always come in line with one caliper point—there is only one easy reading to make.

When folded back out of use the rule is held by a snap-catch.

5 Inch Fay No. 75 outside calipers with quick adjusting spring-nut—4 inch rule graduated in 32nds and 64ths. Send for descriptive folder No. 139. Price, $2.25.

This is one of the many fine tools we make. Get the Starrett 274 page catalog. There is much information in it, which will be useful to you.

Send for complete catalog 19CB.

The L. S. Starrett Company
ATHOL, MASS., U. S. A.

| NEW YORK | LONDON | CHICAGO |
| 150 Chambers Street | 36-37 Upper Thames Street, E. C. | 17 N. Jefferson Street |

Iron Trade Review 1912

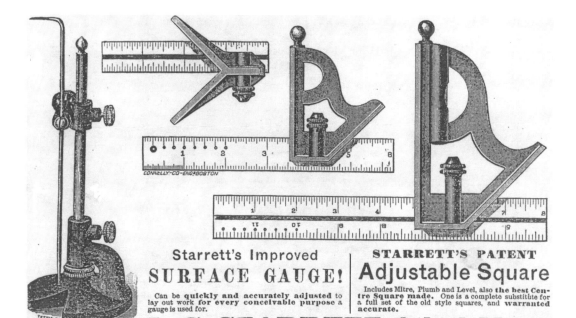

Starrett's Improved
SURFACE GAUGE!

Can be **quickly and accurately adjusted** to lay out work **for every conceivable purpose** a gauge is used for.

STARRETT'S PATENT
Adjustable Square

Includes Mitre, Plumb and Level, also **the best Centre Square made.** One is a complete substitute for a full set of the old style squares, and **warranted accurate.**

L. S. STARRETT, Athol, Mass.
PATENTEE AND SOLE MANUFACTURER.

☞ Send for Catalogue.

Iron Age 1882

Machinery 1897

Thousandths of an Inch are as Important as Feet

when accurate work is necessary. Starrett tools take these thousandths into consideration and give the particular mechanical workman all the accuracy he can use.

That we are constantly working for more accessible accuracy is evident in the addition of

Some New Starrett Tools

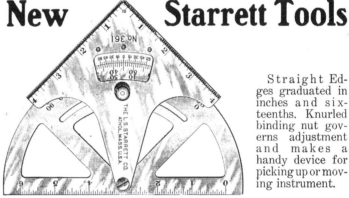

DRAFTSMAN'S PROTRACTOR. No. 361.

No. 361. This high grade Draftsman's Protractor lies flat on the work, is made of sheet steel, reads from left or right with a vernier to read in five minutes.

Straight Edges graduated in inches and sixteenths. Knurled binding nut governs adjustment and makes a handy device for picking up or moving instrument.

Starrett Tool Makers' Calipers and Dividers

No. 277

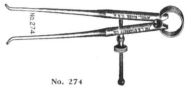

No. 274

No. 560

No. 274 and **277** are two of a new line of calipers and dividers made from round stock with drawn down legs. This makes them hard and stiff. Hardened fulcrum stud, solid nut only, extra strong bows, nicely finished.

No. 560 is an Electrician's pocket screw driver. Handle insulated with hard rubber, ribs insure a firm grip, four blades different widths (3-32″–⅜″) quickly removed from telescopic handle and automatically locked in the end. Spring keeps blades from rattling or losing when cap is off.

Have you the 32-page booklet supplement to Catalogue No. of New Starrett Tools, 18-D? Sent on request.

The L. S. Starrett Co., Athol, Mass.
NEW YORK CHICAGO **LONDON**

Machinery 1910

STEVENS & CO., J., Chicopee Falls, MA

The below advertisement from an 1882 Iron Age magazine includes a previously unrecognized surface gage made by J. STEVENS & CO. The combination square, bevel, straight edge, and drill gage shown is another previously unknown tool recently come to light.

Iron Age 1882

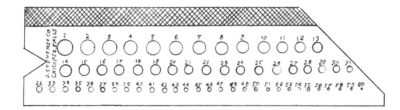

STEVENS-PRENTICE MFG. CO., Milwaukee, WI

Maker, in 1922, of an inclinometer "for the toolroom and inspection departments". The inclinometer was furnished with a vernier scale which allowing reading to 5 min. of arc, with an optional unit available which read to 1 min. The pendulum was furnished with a brake which held it staionary while reading the vernier. Base lengths of 7, 18, and 24 inches were offered.

American Machinist 1922

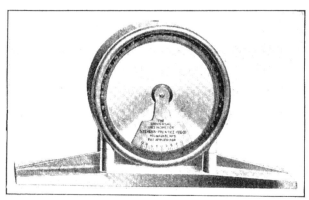

STEVENS-PRENTICE UNIVERSAL INCLINOMETER

STONE & HAZELTON, Boston, MA

A partnership of J.F. Stone and Frederick D. Hazelton. Their primary business was the manufacture of small engine lathes but they were also makers, circa 1876-1882, of the Hazelton patent caliper square. The tool was patented by Frederick D. Hazelton on November 21, 1876.

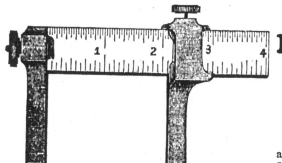

STONE & HAZELTON,
MACHINISTS,
MANUFACTURERS OF
HAZELTON'S PATENT
CALIPER SQUARE,
THE CHEAPEST
and most useful tool of its class in the market, combining scale, square, caliper and depthing tool, and warranted perfectly accurate. Any length of scale furnished. Fine Machine and Model work. Also, Fine Gear Cutting to order.
13 & 15 BOWKER ST., BOSTON.
J. F. STONE.   F. D. HAZELTON.

Worcester City Directory 1877

STRANGE, JOSEPH W., Bangor, ME

See JOHN S. JENNESS.

SULLIVAN, JAMES, Fitchburg, MA

Inventor and maker of dividers, patented March 9, 1880. Inside and outside caliper legs could be substituted for the divider points.

American Machinist 1884

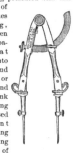

Sullivan's Dividers.

The small engraving presented with this represents a new form of dividers, made by James Sullivan, Fitchburg, Mass. It will be seen that the points are separate pieces, and that they can be swung into position so as to stand vertical to the work, or that one leg may stand in line with the shank while the other is swung at an angle, when used on work of different heights, thus providing for each point standing vertical to the plane of the piece on which it is being used under all conditions of use. The length of the points is also adjustable in the sockets to accommodate irregular surfaces.

The divider points can, if desired, be removed and caliper legs substituted, which can be used, by turning, for either outside or inside calipers; or one caliper leg and one point can be used for lining up work; or one knife-edge can be used for cutting paper washers, etc.

SUTTON, CHARLES D., Kensico, NY

Inventor of registering calipers patented January 3, 1860. The below announcement appeared in *Scientific American* magazine in 1860 but does not make it clear that he was actually making the tools.

SUTTON'S IMPROVED CALIPERS.

The annexed cut illustrates a contrivance for attaching a scale to calipers for the purpose of measuring either the whole or half of the opening or separation of the legs, and thus indicating the diameter or radius of the object embraced by them.

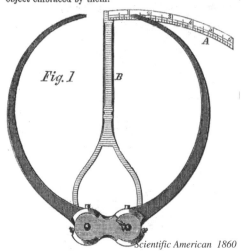

Scientific American 1860

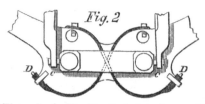

The scale, A, Fig. 1 is attached to the arm, B, which is connected by pivots, $c\ c$, to the fulcrum of the calipers, so that it may be turned out of the way when the implement is applied to an object, and then turned back against the end of the leg to measure the extent of the opening. For the sake of compactness the scale is applied to only one leg, the distance of which it measures from the middle, or point of meeting. To secure the opening of both legs precisely an equal distance from the middle, steel springs or straps are passed around the hubs of the arms, crossing each other in the manner shown in Fig. 2. These straps are provided each with a screw nut, $d\ d$, for drawing them perfectly tight and for adjusting the legs to the middle of the bar, B, before beginning the use of the implement.

TAYLOR & DRURY MFG. CO., Cleveland, OH

Maker, about 1890, of combination inside/outside calipers and dividers nearly identical to the #44 calipers made by the L.S. STARRETT CO. at a later date.

TODT, FRED, Springfield, MA

Maker, in 1917, of a direct-reading, dial caliper. The pointer made a full revolution in $\frac{1}{2}$" and the dial was marked with two scales, one of 32 divisions which would read in 64ths and the other of 500 divisions which would read in thousandths.

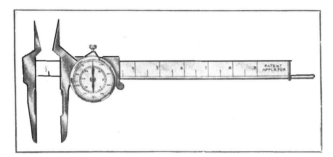

American Machinist 1917

DIRECT-READING VERNIER CALIPER

TURNER & CO., F.L., Orange, MA

Maker of machinist's try squares and depth gages in 1902. The firm also introduced, in 1902, a hub gage which was offered by the L.S. STARRETT CO. in 1903. The Starrett offering, coupled with a complete lack of advertising after 1902, would indicate that the company was short lived. The firm was a reorganization of GREBLE, TURNER & CO., Hamilton, OH, and Orange, MA, which was offering the try squares in 1900.

Identical try squares have been found marked FANEUF & TURNER, ORANGE, MASS.

Machinery, May 1902

MACHINISTS' TOOLS.

F. L. Turner & Co., Orange, Mass., manufacturers of the set of mechanics' try-squares which have already been brought to the attention of our readers, are bringing out several new tools for machinists' use, concerning which they will be pleased to supply information. One of these is a handy caliper, for use in the shop, which will be found useful where it is difficult or impossible to get measurements with calipers of ordinary design. This caliper is termed a hub gage and is shown in the sketch, Fig. 1. Being of small diameter, it can readily pass through the hole in the hub of the pulley or in other work that is to be faced to any required length or thickness, and accurate measurements can thus be obtained

Fig. 1. Hub Gage.

(Additional illustration next page.)

where dependence would usually be placed upon an ordinary steel scale. They also manufacture a depth gage of original design and to supplement their line of try-squares they manufacture an adjustable try-square with ground blade.

Machinery, September 1902

Mechanics' Tools

Accurate Durable Reliable

Turner's Try Squares

Sets of five put up in an oak case, $6.50.

Single squares from 75 cents to $3.50, according to sizes, which range from 2″ to 15″.

24″ and 36″ squares in preparation.

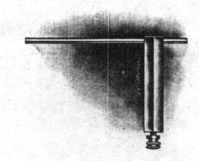

Improved Depth Gauge

2½″ Square with Sliding Blade

This tool for the places where a square with fixed blade could not be used.

Send for circular and price list.

F. L. Turner & Co.
Orange, Mass.

SHOP AGENTS WANTED.

TURNER, WILLIAM D., Providence, RI

Inventor and maker, in 1904, of a moveable pointer which could be installed on a standard micrometer and used to "indicate to the workman the exact line on the graduated sleeve which must register with the zero line when the caliper is set for any given size." Turner was granted a patent for the device on March 8, 1904. The tool is shown in Fig. 1 below. Fig. 2 illustrates a second type which, with a flat end, could be used to show a range within which the measurement should fall. Similar devices are still available.

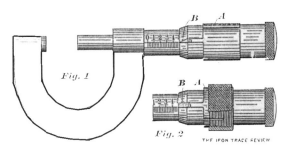

Iron Trade Review 1904

TO FACILITATE THE READING OF THE MICROMETER.

UNION TOOL CHEST WORKS, Rochester, NY

Maker of machinist's tool chests in 1916. The company later operated as the UNION TOOL CHEST CO., INC.

Machinery 1916

UNIQUE

Name marked on an English-made indicator attachment for calipers. The indicator was patented in the U.S. February 23, 1915, by John H. Grant and was imported by the William H. Simpson Co. There is a very similar American-made tool marked FINDER.

VEEDER-ROOT CO., Hartford, CT

Successor to the VEEDER MFG. CO. about 1928. As late as 1960, the company continued to make the line of small speed indicators which had been made by the VEEDER MFG. CO. since 1908. In 1929, they took over the speed indicator line which had been made by the CORBIN SCREW CO. This tool, shown below, was a true tachometer, reading in RPM.

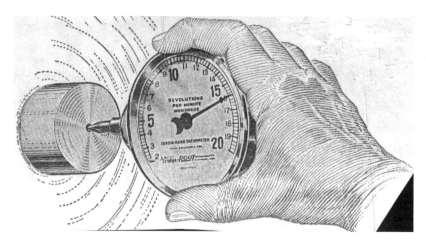

American Machinist 1929

VERWYS TOOL CO., Grand Rapids, MI

Maker, in 1919, of Verwys' adjustable offset center head for attachment to a combination square blade. The tool was patented March 11, 1919, by Abraham Verwys.

American Machinist 1919

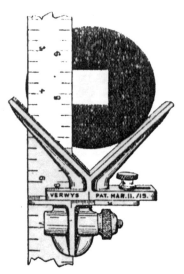

VERWYS' ADJUSTABLE OFFSET CENTER HEAD

VICTOR RULE CO., Hartford, CT

A 3" machinist's rule marked as above has been reported. Date is uncertain but it appears to be of the 1870-1880 period.

VOLIS PRECISION TOOL CO., Detroit, MI

Operated by Harry Volis, the firm was the maker, about 1921-1926, of bore gages patented by Volis on May 23, 1922, October 20, 1925, and August 10, 1926.

VON CLEFT & CO., New York, NY

Maker of one piece, forged calipers very similar to those made by J. STEVENS & CO. Period of manufacture is uncertain but probably about 1870-1880.

VOSE, A.A., address unknown

Maker of a surface gage finely adjustable by a worm and wheel mechanism. The gage was patented January 15, 1889, by Carl G. Osteman. The only known example is marked:

PAT'D JAN 15 1889
MANFD by A.A. VOSE

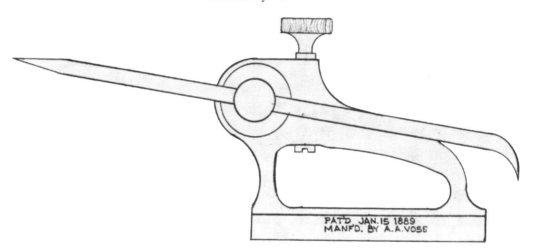

WALDEN TOOL CO., Boston, MA

Maker of worm and gear wheel fine adjusting calipers marked PAT APD FOR. No patent has been located. Founded by Frederick E. Walden after selling the Walden Mfg. Company in 1907.

WALKER TOOL CO., Milwaukee, WI

See entry in *Makers of American Machinist's Tools,* first volume, and the HENNECKE-WALKER CO. in this volume.

WALTHAM DIAL GAUGE CO., Waltham, MA

Introduced, in 1925, its 4" diameter "Giant" dial indicator which was probably the largest production dial indicator ever made. Its advertising slogan was "Throw Your Specs Away!"

WARWICK TOOL CO., Middletown, CT

As noted in *Makers of American Machinist's Tools,* first volume, this company was one of the first commercial makers of surface gages, probably starting during the Civil War. An 1867 flyer offers drill chucks, surface gages, hand vises, hacksaw frames and Parker's gear-cutting attachment, patented July 3, 1866. As shown below, the surface gage was also sold through A.J. WILKINSON & CO. Several examples are known.

All Warwick's products, except the gear-cutting attachment, were offered by the HUBBARD & CURTISS MFG. CO., Middletown, CT, in its 1872 catalog. It appears that Warwick was absorbed by Hubbard & Curtiss when it incorporated in April, 1872.

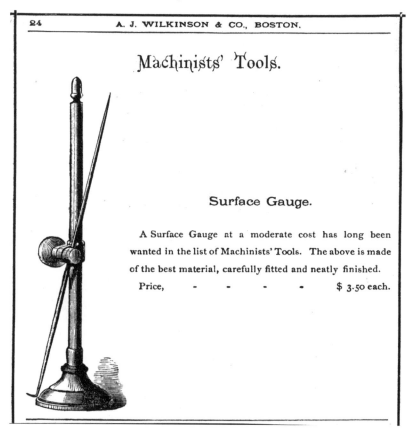

WASHBURN, F.O., Millville, MA

Inventor of machinist's calipers patented January 10, 1865. The only specimen examined is unmarked so it is unclear if Washburn was also the maker. *(Illustration on next page.)*

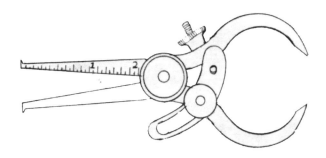

WEDELL & BOERS, Detroit, MI

Makers of machinist's tool cases ca. 1910-1916. Its 1910 ad notes that the box is "covered with genuine cowhide leather," an unusual feature. Drawer fronts were solid mahogany.

Machinery 1916

WEISSENBORN, EDWARD, Jersey City, NJ

Inventor and maker of calipers and dividers for which he was granted a patent May 4, 1886. They appear to have been made for both machinists and draftsmen.

WELCH, T.F. & CO., Boston, MA

Maker of twist drill gages and tap wrenches in 1893. See entry in *Makers of American Machinist's Tools, first volume* for products in 1902.

(Illustrations on next page.)

Twist Drill Gauge.

We illustrate herewith a new twist drill gauge, the opposite sides being shown in Figs. 1 and 2, manufactured by T. F. Welch & Co., 65 Sudbury street, Boston, Mass. The tool is made of cast steel, hardened. The holes, it is stated, are gauged accurately, and the principal improvement embodied in this tool is that it combines a drill gauge with the tap

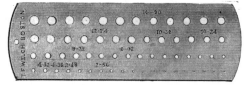

Fig. 1.—*Twist Drill Gauge.*

drill sizes. This is accomplished by indicating on reverse side the tap drill sizes by means of numbers. It is claimed that this is a great advantage over the old

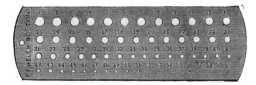

Fig. 2.—*Gauge for Taps.*

method of guessing the sizes required by the different taps. The point is made that the tool is well made and finished and sold at a moderate price.

The accompanying illustration is of a tap wrench being put upon the market by T. F. Welch & Co., 65 Sudbury street, Boston, Mass. It is made of steel, and

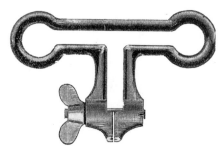

The Welch Tap Wrench.

the jaws are regulated by a thumb screw. The wrench is intended for small size taps up to ¼ inch, and the manufacturers claim for it effectiveness and cheapness.

Iron Age 1893

WELLES, F.A., Milwaukee, WI

The September, 1892, F.A. WELLES catalog, issued by tool dealer F.L. Stoddard, Minneapolis, MN, gives us a full description of the early Welles products and an interesting treatise on the art of calipering and fitting as practiced in the late 19th century.

This is the earliest Welles catalog now known and makes clear that the two types of surface gages and the Welles & Harrison patent trammel were, in fact, offered for sale. The calipers are well known, at least in the mid-west, but no examples of the surface gages or the trammel have been seen to date. This has led to doubt, now resolved, that the tools ever got beyond the press release stage. (Since the above was written, a Welles & Harrison trammel has surfaced at a Wisconsin flea market.)

For more details on F.A. WELLES and the later WELLES CALIPER CO., see *Makers of American Machinist's Tools,* first volume.

(Catalog is illustrated on the following eight pages.)

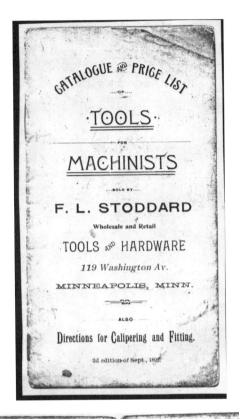

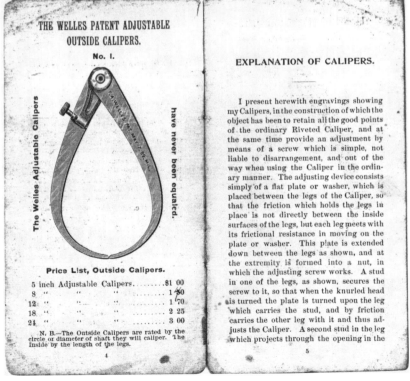

plate, prevents the latter from being moved too far by the screw, as, of course, only a very little movement is ever needed.

It will be seen that by this construction the Caliper can be adjusted in whole or in part by the ordinary method, **without having to stop to loosen or tighten anything**, and that one adjustment in no way interferes with the other.

These Calipers have points of excellence that are not found in any others on the market; the design, proportion and manner of construction being based on experience in the use of Calipers.

Mechanics skilled in fitting with Calipers will find it to their interest to give the Adjustable Calipers a trial.

All Calipers are warranted to be RELIABLE.

In localities where my tools are not sold by the trade, I will send any tool, postpaid, on receipt of price.

THE WELLES PATENT ADJUSTABLE INSIDE CALIPER.
No. 2.

The Welles Adjustable Calipers have never been equaled.

Price List, Inside Calipers.

5 inch Adjustable Caliper.........$1 00
8   "   "   " 1 40
12 "   "   " 1 70
18 "   "   " 2 25
24 "   "   " 3 00

N. B.—The Outside Calipers are rated by the circle or diameter of shaft they will caliper. The Inside by the length of the legs.

THE WELLES FIRM JOINT CALIPERS.
These Calipers are very Popular.
No. 3.

Price List, Outside Calipers.

5 inch Firm Joint Caliper.........$ 60
8   "   "   "   80
12 "   "   " 1 00
18 "   "   " 1 75
24 "   "   " 2 40

N. B.—The Outside Calipers are rated by the circle or diameter of shaft they will caliper. The Inside by the length of the legs.

THE WELLES FIRM JOINT CALIPERS.
These Calipers are very Popular.
No. 4.

Price List, Inside Calipers.

5 inch Firm Joint Caliper.........$ 60
8   "   "   "   80
12 "   "   " 1 00
18 "   "   " 1 75
24 "   "   " 2 40

N. B.—The Outside Calipers are rated by the circle or diameter of shaft they will caliper. The Inside by the length of the legs.

THE WELLES FIRM JOINT MORFADITE.

Those that have used these Morfadites find that the point is PROPERLY HARDENED.

No. 11.

Price List.

5 inch............................60 cents
8 " ....80 "

N. B.—The Morfadites are measured by the length of the legs.

THE WELLES FIRM JOINT DIVIDER.

This is a new tool. It is likely to prove very popular. Legs are stiff and points are PROPERLY HARDENED.

No. 12.

Price List.

5 inch............................60 cents
8 " ....80 "

N. B.—The dividers are measured by the length of the legs.

THE WELLES PATENT ADJUSTABLE DIVIDER.

This is also a new tool. It is not necessary to loosen or fasten anything. It is always ready.

No. 13.

Price List.

5 inch Adjustable Divider..........$1 00
8 "   "   " 1 40

N. B.—The dividers are measured by the length of the legs.

THE WELLES PATENT ADJUSTABLE MORFADITE.

New tool. Will be found very useful.

No. 14.

Price List.

5 inch Adjustable Morfadite........$1 00
8 "   "   " 1 40

N. B.—The Morfadites are measured by the length of the legs.

THE WELLES PATENT SURFACE GAUGE.

Style No. 1.

(Pat. Jan. 20, '91.)

10 inches usual height, 18 inches when extended.

Price, style No. 1...... ....$3 75

THIS Gauge has features that will commend itself to practical mechanics. The standard can be swiveled to any position and then firmly clamped. This is accomplished by the revoluble standard supporting wheel. The wheel is firmly clamped by a lever screw forcing a block against the side of the wheel near the edge. When at the usual height the standard is about 10 inches high. For any height over 10 inches the telescope tube is raised and then clamped by a lever screw. The standard is about 18 inches high when the tube is raised to the full height.

The gauge pins (they look like screw heads in the cut) are round pieces of steel fitted to reamed holes in the base. The upper half of them is slotted, thus causing a slight friction. When the gauge pins are needed they are pushed down with the fingers and they will stay in any position they are placed. The needle is swung by a lever, between the needle collar and sleeve. The lever is swung by a knurled thumb screw, which is journaled in a bracket extending from the sleeve. (The principle of adjustment is the same as the WELLES ADJUSTABLE CALIPERS.)

The sleeve is so cut as to clamp sufficiently on tubing and the rest of the pressure caused by needle bolt is taken by solid metal of the sleeve without springing the tube.

Below are a few uses of the gauge:

The standard may be inclined to go under the crosshead of any planer. When space on planer platen is limited the standard may be placed in any position to accommodate the work. On account of limited room on milling machine platens this gauge is invaluable.

Work may be lined for a side cut by use of the gauge pins and swiveled standard, etc., etc.

This gauge can be used anywhere the ordinary gauge can, and can be used in places where no other surface gauge will go or reach.

It is impossible to mention all of the good points of the gauge in this description. The aim of the designer being to make a practical and complete tool such as actual experience can dictate.

THE WELLES PATENT SURFACE GAUGE.

Style No. 2.

This Gauge is similar to No. 1, except that it has no extension tube.

(Pat. Jan. 20, '91.)

12 inch Gauge.

Price, No. 2........................$2 75

THE WELLES & HARRISON'S PATENT BAR CALIPER AND TRAM GAUGE.

Following is a description of the tool:

A round steel bar, on which there is a sliding tube for extension. The bar has a V groove on the lower side of its entire length, likewise the tube (which is of very stiff brass), has an indented V groove its entire length to match the bar. A clamping collar is fitted on the end of tube and then sweated, practically making it a part of the tube; both the end of tube and the collar are then cut to allow the clamping screw to clamp the tube in any position on the bar. Both heads have means to prevent rotary motion. The head that fits on the end of the bar is a snug sliding fit, but the screw has power enough to move it for fine adjustment. This obviates the necessity of loosening or tightening a set screw before and after an adjustment is made. A small set screw (not shown) permits additional friction, by forcing a block into the V groove. This head has a long pointer which may be raised or lowered to suit the work. A slot is made in the bar to allow the pointer to pass through it and to permit endwise movement of the head while the pointer is in position. The head which slides on the tube is split and has a clamp screw which will hold it firmly. Both heads have hardened steel caliper faces, which will caliper from 0 to nearly 20 inches. In use, the adjusting screw should be at the right hand.

The length of this tool when closed is about 12 inches, but it will scribe a 40-inch circle when extended. Thus when a small caliper or tram is needed, the bar will be short and not in the way, and when a large caliper or tram is wanted, it will only be necessary to extend and clamp the tube. When put away in the tool box it will take but little room.

The bent end of the pointer allows this tool to be used as a "morfadite."

Price........................... $3 50

Pat. May. 20, 1890.

DIRECTIONS FOR CALIPERING AND FITTING.

COPYRIGHT 1892, BY FRED A. WELLES.

Observing that very little has been written on calipering, it will not be out of place to give a few hints for the benefit of those learning the machinist's trade. I do not wish to be understood that the methods described are **the way**, and **the only way**, of using calipers properly; but as a way that is practical and gives good results.

FITTING AND CALIPERING IN A LATHE.

There are three general methods of fitting:

1. By fitting exclusively with calipers.
2. By trial and following the trial with calipers.
3. By cut and try.

The **first** is a method that is very valuable, as it often saves hard work, as well as being the quickest way. Take large shops for instance, the lathes are often quite a distance from the machines being built; thus, if two journeys — one to set calipers and one to drive stud or bolt in place — will finish the job, a great deal of sole leather and extra work is saved. If the bolt is too large it must be driven out, re-filed and perhaps re-turned — depends upon operator. This is a kind of unnecessary labor in which there is no satisfaction. Aside from saving time and hard work, the first method has a feature that is worth while striving for; the satisfaction of driving a bolt in at first trial can best be appreciated by those who can do it. As an example, suppose the journeyman machinist is in a new shop (to him); his first job may be a stud or bolt which is to drive into the frame of an engine. He may be capable of making fits the first time, and he may not. If he is capable and drives the stud in at the first trial, it will not take long for those who are watching this new man to see that he is a mechanic. But those that cannot make caliper fits are not necessarily poor mechanics.

The **second** method is necessary in certain kinds of fitting. For beginners, it is the best way to start; but, even then it would be a very good idea to have two sets of calipers in use — (four calipers). Set the inside calipers to the hole, then match the outside calipers to the inside. Both calipers should be set with great care. Then match another pair of outside calipers to another pair of inside calipers. The last pair of outsides are to be used in turning to size. The first mentioned set of calipers are to be put away where they will not be disturbed. Turn the shaft, if it be such, so the end of it will just enter the hole, then twist it a little to show a bearing. Now set the calipers to the end of the shaft, so they just touch where it is brightened by contact with the hole. File the brightened part and the rest of the shaft that it is to fit, so that the calipers will not quite touch; this is for a close working fit. If the hole is straight and parallel, it will not require many trials to fit it. After the shaft is fitted, get the first set of calipers out and see how the outside calipers compare with the shaft. If they just touch the shaft it shows well. If they do not touch the shaft, look at the hole to see if the shaft brightened it uniformly. If the hole was bored in the lathe, without reaming, it may show slight irregularities, or chatters, that were not noticed until working the shaft in the hole.

The **third** method is adapted for taper and not parallel fittings.

CALIPERING HOLES.

Reamed holes are usually the easiest to caliper and fit, as they are more likely to be round and parallel. When a job has been bored in the lathe and not reamed, more care should be taken, as the hole may not be parallel and round.

To use inside calipers it is usually best to hold one of the toes against the wall of the hole with the index finger of the left hand, the thumb, index and second finger of the right hand supporting the caliper at the joint. Thus one toe will be stationary and the other free to move.

The fitter will sometimes be obliged to fit a shaft in a hole that is not round, at the same time it may appear round when tested with the calipers. With a long, springy boring tool, particularly if the job is hurried, odd shaped holes are likely to be the result. Three and five-cornered holes, so-called because the circle has a very slight shape of a triangle or pentagon, are deceptive to the calipers. Thus it should be understood that the error tends to make the shaft too large. But, whatever the shape of the section of the hole is, the inside calipers, if good ones, can **never** give a size that is **smaller** than the **round** shaft that fits it.

In using inside calipers it is better to set the calipers a trifle smaller than the hole, (except in driving fits.) Set them so the moving or vibrating toe will all but touch when the toes are exactly opposite in the hole.

CALIPERING SHAFTS.

The outside calipers may be held in one hand for calipering shafts that are less than three inches in diameter. But for shafts over eight inches in diameter it is better to have the assistance of the index finger of the left hand in holding one toe against the shaft, and allow the other toe to be swung or vibrated in an arc in both directions, the caliper being supported near the joint by the fingers of the right hand. The reason for swinging or vibrating one toe is to find the opposite part of the shaft. Allow the toe to touch very lightly when the opposite part of the shaft is found.

MATCHING CALIPERS AND FITTING.

After the user is able to use both outside and inside calipers with a very light touch, then fitting exclusively with calipers can be attempted. Do not be afraid of using calipers too lightly, as calipering lightly is one of the most important points. Take **time** in setting calipers to be **sure** that they are **right.** If you cannot take time to caliper with care, all I can say is, don't depend on the calipers entirely.

But, as a rule, it will save time to set calipers properly, even on a job as small as fitting a three-quarter inch bolt.

Generally, short holes require the shaft to be a trifle larger than for long holes.

Holes that are simply bored in the lathe and not reamed are often harder to fit than those that are reamed; the boring tool often produces slight irregularities that are bothersome to the fitter.

Long holes that are drilled and reamed in the drill press, often give trouble in fitting, as they sometimes take a curved course, caused by the drill meeting a blow hole or any other obstacle that will deflect it from a straight course.

Confidence is essential in making caliper fits, so do not attempt to make fits unless you have good, reliable calipers. If such fitting is attempted and the result is a bad fit, on account of using poor, unreliable calipers, the confidence in the ability to use calipers is shaken.

Fitting requires a great deal of care, even when the best calipers are used. Poor ones can never give satisfactory results.

In matching the outside to the inside, in small calipers; the thumb and index finger of the left hand hold the lower leg of the inside caliper, the second finger of the same hand serving as a guide to hold the lower toe of inside and lower toe of outside together in their proper position.

The right hand is used to support the outside caliper at the joint and to cause the upper toe of outside to move or vibrate so as to just touch the upper toe of the inside. The right hand caliper is the one to be adjusted.

In matching the **inside** to the outside caliper the outside is held by the thumb and fingers of the left hand; the index finger of the same hand serves as a guide at the junction of the lower toes, otherwise the operation is the same as described above.

In matching 12-inch and larger calipers it is often easier to lay one caliper flat on the bench, or lathe, and allowing the toes to project. The other caliper being held in the right hand, the thumb and index finger of the left hand serves to hold the stationary toes together, the other toe of the right hand caliper is the one to be vibrated.

The main object, when matching, is to hold the calipers lightly, and without springing the legs, and above all, to make the **inside**, at its **largest**, touch lightly the **outside**, at its **smallest**. By practice and perseverance it can be accomplished with certainty.

I will give rules that I have found to be of great aid in fitting:

For a **driving fit,** have the toes of the insides just touch the opposite walls of the hole, match the outsides to the insides. Make the shaft so that the toes of the outsides will touch it lightly.

For an iron to iron, or snug working fit, set the insides a trifle smaller than the hole, so that when one toe is held stationary the other toe will touch only when vibrated a distance, the measurement is on the circle and the toe is **not** to touch in the short arc. The amount of circular vibration allowed is according to the size of the hole. See table below:

TABLE OF VIBRATION FOR INSIDE CALIPERS FOR SNUG WORKING FITS.

Circular vibration for a 2½ in. hole, 3-16 inch.
" " " 4 " " ¼ "
" " " 9 " " ⅜ "
" " " 12 " " ½ "
" " " 16 " " ⅝ "
" " " 18 " " ¾ "

After the inside calipers are set with the correct amount of vibration, match the outsides to them, then make the shaft to size so that the outsides will lightly touch the shaft.

These rules are given for short holes, say, length equal to the diameter.

In driving fits there are a good many conditions to be taken in consideration, such as tightness of fit required, the kind of metal used in the shaft or surrounding the hole and thickness of metal surrounding the hole, etc. etc.

The above rule for driving fits is adapted for a hole in cast iron length equal to the diameter, plenty of iron around the hole and an iron or machine steel mandrel to fit the hole. The fit will be snug enough to stand a fair size cut, the mandrel being driven by a dog when turning in the lathe.

The fitter should be thoroughly acquainted with the hole in which the shaft is to be fitted. If the hole is tapered or out of round or both tapered and out of round, it will be necessary to do some trying, but always follow the trial with the calipers and see how the part of the shaft that shows a bearing corresponds with the rest of the shaft.

In some cases where a very close working fit is wanted, it is better to turn the shaft to a light driving fit, then oil the shaft and sprinkle a little plumbago on the oil. Drive the shaft into the hole with a block of wood, then turn it around in the hole with a dog. Take the shaft out and see where it bears. Next put the shaft in the lathe and polish the bright spots with No. 0 emery cloth; give end motions when polishing, thus the scratches will cross each other. Wipe off all the emery, again insert it in the hole and turn with the dog as before, and if it does not work to a fit polish bright parts again, etc.

The oil and plumbago are used the first time to prevent the shaft or hole from cutting.

This method is especially advantageous where the whole has been bored and not reamed. The hole is improved by the lapping and a closer fit can be made.

The shaft in this case we will suppose is machine steel, the last cut or two is made with a square-nosed tool, using a fine feed and a fast speed; use no oil. With a good lathe, set to turn straight, no file will be needed (the shaft is supposed to be short).

Shafts that are for a free running fit need to be a trifle smaller. Either set the inside calipers for a driving or working fit; then place a piece of paper the correct thickness between the work and the outside caliper's lower toe.

In using outside caliper on planer work, it is often best to hold one toe of the caliper against the work and vibrate the other leg to find the exact opposite side. Have the vibrations as small as possible without having the leg touch when the toes are in opposite positions.

CALIPERING PLANER WORK.

In making an iron to iron fit that requires pounding with a block of wood to drive it to place, the following way will answer the purpose in most cases: Set the insides so as to just touch the opposite walls, match the outsides to the insides. Plane the work so that the outside caliper will not quite touch when the toes are opposite, allowing as little vibration as possible.

If the work is not to be an iron to iron fit, paper the proper thickness placed between one toe and the work will accomplish the object.

WHITMORE, ANDREW E., Boston, MA

Inventor and maker of a micrometer adjusting, caliper gage patented July 27, 1869. The only known specimen is marked:

A.E. WHITMORE
PATD JULY 27, 1869

The gage appears in the A.J. WILKINSON & CO. catalog of 1875 as shown below. Note the $10.00 price tag; a lot of money for a machinist making $2.00 a day or less. Whitmore also received patents for a similar gage on April 5, and July 5, 1870, and for an odd looking micrometer on May 24, 1887. No tools made under these three patents have been observed.

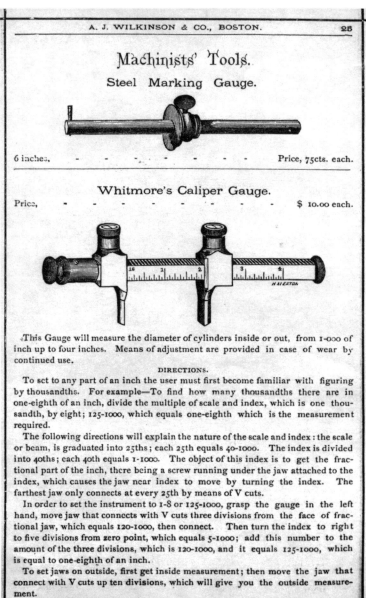

WIET-GOETHE,   Sacramento, CA

Maker of a combination gage patented March 21, 1905, by Eugene Wiet. When the entry for *Makers of American Machinist's Tools,* first volume, was written, no complete tool had been examined. This is still the case; even though several other examples have been found, none have been complete. The tool is poorly made of light sheet metal and with several loose pieces easily lost. The below 1905 advertisement, offering the tool free with a $1.00 magazine subscription, gives us a look at the complete tool. Note the claim "made by first class mechanics and no expense is spared to insure accuracy."

FREE to DRAUGHTSMEN! ELECTRICIANS! TOOL MAKERS! PATTERN MAKERS! MACHiNISTS! EVERYBODY!

A new handy and convenient device for taking measurement is the **WIET-GOETHE COMBINATION GAUGE**

Full Size

NOW offered for the FIRST TIME to those interested.

The Scale is graduated to 64ths of an inch. The Protractor in degrees and numbered.

DID IT ever occur to you when called upon to perform some mechanical work to be confronted by almost insurmountable difficulties because you did not have handy the right kind of a tool to take measurements? ¶ Is it practical for any man to carry a cumbersome tool chest to take measurements? ¶ Suppose you could carry in your vest pocket all in one instrument

AN OUTSIDE CALIPER	A STRAIGHT EDGE	A TRY SQUARE
AN INSIDE CALIPER	AN ANGLE GAUGE	A CENTER SQUARE
A DIVIDER	A DEPTH GAUGE	AN ANGLE PROTRACTOR
	A CENTER GAUGE	

and many other combinations too numerous to mention, but understood and appreciated at once by all having to deal with any kind of measurements. ¶ We now present a tool which will enable you to fill all these requirements and take all measurements, within its scope, accurately and quickly. ¶ It is made of superior steel, carefully and thoroughly tested as to its accuracy. ¶ It will be found to be a novel and valuable addition to the tool chest of any mechanic. ¶ The practical men will find constant and novel uses for it under any and all circumstances. ¶ It will enable one to solve all the different and difficult problems which arise in the shop or elsewhere. ¶ It is so portable and convenient; being small enough for the pocket—that it can always be at hand. ¶ The Gauge is made by first class mechanics and no expense is spared to insure accuracy. We commend it as a finished and perfect tool for the various uses to which it can be put. ¶ The Combination Gauge is enclosed in a neat leatherette case with metal trimmings, so as to be conveniently carried in the pocket. ¶ You can get one of these invaluable Gauges **free** if you accept the proposition outlined below.

AGENTS WANTED

WIGHT, HIRAM P., Worcester, MA
See COPELAND & CHAMBERLIN

WOLFE, J.L., Bridgeport, CT   later:
WOLFE & SON, J.L., Bridgeport, CT

Maker, in 1909, of a machinist's test indicator with an unusually wide range of 1/16" in either direction from the center zero point. The tool is marked PAT APPLD FOR.

In 1912, the firm, now operating as J.L. Wolfe & Son, offered a modified version as the thread test indicator shown below. This was one of the earliest to display lead error through a two inch span. Note that the indicator was still marked PAT APPLD FOR. Joseph L. Wolfe finally received a patent April 6, 1915, nearly seven years after his application.

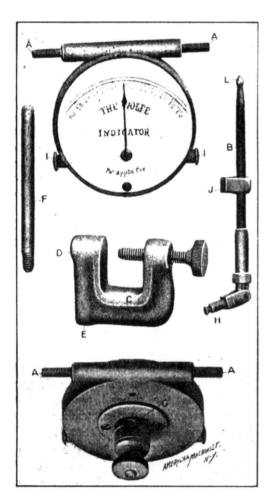

TEST INDICATOR

American Machinist 1909

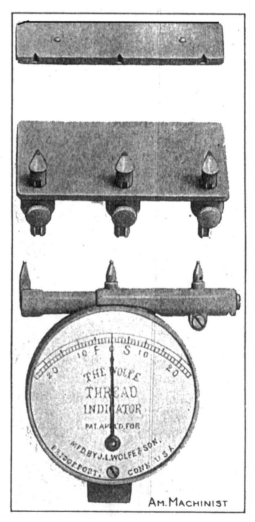

FIG. 1. THREAD INDICATOR AND BALL POINT AND CENTERING GAGES

American Machinist 1912

WOODWORTH SPECIALTIES CORP., Binghamton, NY

Maker, in 1949, of indicating calipers made of sheet stainless steel formed into hollow legs, one of which contains the indicator arm. The specimen examined is marked PAT. APPLIED FOR, but no patent has been found.

WORCESTER RULE CO., Worcester, MA

Maker of the Long patent gear-calculating rule in 1866. See LONG, CHARLES B.

YORK CO., S.M., Cleveland, OH

This firm was listed in *Makers of American Machinist's Tools,* first volume, as a maker of combination inside/outside calipers. We now find that S. Milton York (1848-1898) was a Cleveland machinery dealer operating as S.M. YORK CO. from ca.1890 to 1898. It is likely that calipers marked with the York name were advertising pieces and were actually made by the TAYLOR & DRURY MFG. CO., Cleveland, OH.

York was also the patentee, September 15, 1874, of a very useful looking combination protractor, depth gage, square, etc. (See *American Machinists' Tools, an Illustrated Directory of Patents*, page 190) None of these tools have been observed.

YOUNG-FISHER INCLINOMETER CO., Milwaukee, WI

Maker, in 1921, of a machine shop inclinometer with one dial reading in degrees and a second dial, geared 1 to 36 to the first, which reads in minutes. The company claimed that the tool was superior to the sine-bar for the accurate inspection of angles.

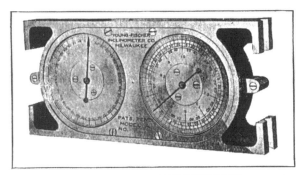

American Machinist 1921

YOUNG-FISCHER INCLINOMETER

APPENDIX I

THE DEVELOPMENT OF FIXED CALIPER GAGES, 1862-1878

The following is excerpted from an 1896 lecture delivered at Stanford University by John Richards and printed in *American Machinist* magazine of July 16, 1896. Richards was a developer of the fixed caliper gage, the founder of the American Standard Gage Works and a noted authority on precison measuring. Comments in brackets have been added by the author.

The true standard in this country is the meter, legalized in July, 1866, decimally divided downward to the millimeter, and with multiples upward the same, but by custom, and following the British measures, our machine work is mainly with measures corresponding to divisions of a yard, with divisions downward of three and twelve, stopping at an inch, which is twenty times too large for convenience, and going up by factors of 5 1/2, 40 and 8 to a mile, all of which is as absurd as it is inconvenient.

In 1860, when my connection with this matter began, there were no gaging implements in this country, except Whitworth pins and collars [plug and ring gages], imported at an expense of $300 to $600 a set, according to the number of sizes embraced, and these gages were found only in a few of the largest shops. Such gages are only for reference, from which other implements are made, hence did not meet the real want of gages that could be given to the men for practical use. Some firms had made forged caliper gages with round or curved contact points, but there was no manufacture of standard gages in this country, nor was there in England except of pins and collars, down to 1878.

With this much in respect to the derivation and nature of gaging implements, I will now revert to the history of their manufacture in this country, in so far as I had a part in this matter.

In 1862, when foreman in the works of the Ohio Tool Co., at Columbus, Ohio, I had occasion to make some duplicate works, and not having any means of attaining standard sizes, I sent to Jones & Laughlin's, at Pittsburg, makers of cold rolled shafting, asking them to send a piece of 2 inch shafting that would fit the Whitworth collars they used in their rolling processes. A selected piece of this shaft about one foot in length, two inches diameter, became a standard, from which was derived a series of sizes from one-half to three inches, in the following manner.

The cross-feed screw of one of the engine lathes was measured with a rule, and found to be tolerably accurate, the pitch being eight threads per inch. A conical pin of cast iron was put in this lathe and turned in steps by eights and sixteenths of an inch, but was left a little larger. Then a finely tempered square tool was put in the lathe and its edge set true or parallel to the work. The cross-feed screw was provided with a disk divided into four parts, with a detent that would lock it at these four points. The two inch step on the come was then turned, or scraped until it would caliper the same as the piece of shafting before mentioned. Then the tool was advanced a quarter or half turn and fed over the next step, and so on down to the half-inch size.

The same process was gone through from the 2 inch step upward to the 3-inch one. This cone was mounted in an inclined position on a cast iron base and became a standard for sizes, resting, it is true, on a very uncertain standard, but much better than no standard at all. [This standard is clearly illustrated in Richard's patent of October 15, 1867. See *American Machinist's Tools, An Illustrated Directory of Patents,* page 172.]

It was soon discovered that fixed calipers were required, and I proceeded to make these by cutting out lune-formed pieces from sheets of cast steel, also had some made from forgings of cast steel. [These fixed calipers were also patented by Richards on October 15, 1867.] A set of these gages, about forty in number, was sent to Messrs Brown & Sharpe, at Providence, R.I., to be ground to size, with flat points, and I believe were the first gages made in this manner. Previously the points had been made curved or convex, offering only line contact to withstand wear. A set of these gages were taken to Messrs. J.A. Fay & Co.'s works at Cincinnati, Ohio, in 1865, and are no doubt in use there at this time.

Patents were taken out on the lune-formed, flat point calipers, and on the corrective cone, in 1867, but various circumstances prevented farther work on gages until 1869, when I went to Philadelphia to found the making of my gages and standards.

Everyone regarded the matter of such gages as a mystery, and set small value on the audacity of a young man from Ohio, who proposed an innovation of the Whitworth system. I learned, as afterwards proved almost true, that the making of gages was in some respects a mystery, and that no one knew how the processes were carried out, but the main impediment was that I proposed a new system. [Not much has changed in the last 128 years.]

No one would aid in such an undertaking, and I went on to England, where the conclusion was that the means I had at command would be of no use in founding such a business and the scheme was abandoned until 1872. Then having more means I rented an office in Philadelphia, and arranged to have room in the basement to carry on the manufacture of gages. [This was the beginning of the American Standard Gage Works.] In this office were prepared the designs and detail drawings for ten special machines adapted to the various processes required in making calipers and gages.

A contract was made with Baxter D. Whitney, of Winchendon, Mass., to construct these machines, which he did with an accuracy that it would be hard to excel at this day. The work occupied about a year, during which time I was in England, and then came the great panic of 1874, which stopped all industry. No one could conceive of what was to come; shops were closed, and skilled mechanics had to beg for their bread.

My gage making machines were coated with lead and tallow, boxed up carefully, stored away at Winchendon until August, 1877, when they were sent to Philadelphia, set up, and the gage works were founded fifteen years after the first outfit of gages was made at Columbus, Ohio.

In 1875 or 1876, Prof. John E. Sweet at Cornell University began and carried out some very complete experiments in making standard calipers that deserve notice in any history of the art in this country. Some very accurate caliper gages made at the university were exhibited at the Centennial Exhibition of 1876. But Prof. Sweet, in talking of them, said: "We will not make any more of these calipers."

In 1877, it was thought that in a month of so calipers and corrective gages would be ready for sale, but these expectations were foiled by several circumstances, principal among which was the failure of the measuring machine, fitted with graduating screws made at the Whitworth Company's works in England, and guaranteed as to accuracy.

This matter had before been thought of, but there was scarcely a doubt that the pitch of the screws was correct. Subsequent experiments, however, proved that not only the aggregate pitch was wrong, and the screws not parallel, but the relative pitch in so short a length as seven inches would not do to depend upon. The screws were, as the Whitworth Company afterwards maintained, carefully made, but the delicacy of measuring tests demands more accuracy than can be attained by the pitch or movement of screws.

The first experiment in measuring was made by preparing six rods, the points nicely finished, and the central part covered with several layers of thick soft paper to prevent induction. [Expansion due to heating when handling.] These rods were carefully fitted into the machine when set at six inches, temperature and other conditions being carefully observed. The rods were then uncovered and fitted into a groove cut on the side of a bar of pine wood, and taken to Messr. W.B. Bement & Sons' works, where a Whitworth master screw, with the necessary conveniences for testing the rods, was supplied, and the experiments conducted by the firm. By careful comparison it was found that the measuring machine in comparison with the screw recorded about one in ten thousand short. [Richards was struggling with the basic problem of locating a standard to use when making accurate gages. Note, however, how far he had progressed since 1862.] This, of course, stopped gage-making for the time. The test rods were then taken to London, and in three separate experiments on different standards, two by myself and one by Gen. B.C. Tilghman, it was found that the rods were short, the variation being but little in the three cases, and coresponding very nearly with the less perfect experiment at Philadelphia.

The next operation was to procure four standard test rods, from Troughton & Sims, of London, adjusted to the imperial yard of Great Britain and its divisions, by which the measuring machine in Philadelphia could be adjusted. The adjustment of the measuring machine occupied one month's time. The room was kept at one temperature, as nearly as possible, and thousands of reading were taken off, and when finally done, and the screws calibrated, the profile of error was curious to observe; the profile of one screw was convex, so to speak, and the other concave in respect to the axes.

When finally we were prepared to adjust gaging implements there arose another impediment, more formidable still. The contact points of the calipers with flat faces could not be ground parallel. The grinding wheels have to rotate parallel to these faces, and pass over them, and as the pressure is as the area of contact, the unavoidable elasticity in the wheels and their mountings produced a curved surface.

I remembered then Prof. Sweet's significant remark, and we applied to him for information. It was as expected. This was the difficulty he had met with. He kindly came to Philadelphia to aid us with his own experience and suggestions, but there seemed no way out of this dilemma, until one day light came from an unexpected quarter.

Mr. J. Morton Poole, Sr., of Messrs. Poole & Sons, Wilmington, Del., came into the works to examine the gage-making process. At this time I imagine that no one in the world had given so much attention to grinding as Mr. Poole. He had a laboratory for experiments and the making of grinding wheels. He had collected carborundum and other abrasive mineral from all parts of the world. Some of his grinding operations on paper calendering rollers were marvelous, and so intricate as to almost defy explanation. He would grind piles of these rollers eleven in height so as to exclude light between them, and the line of light, as it is called, is commonly estimated at the ten thousandth part of an inch. [I have never heard the term "line of light" before, but the principle is commonly used by skilled machinists to this day.]

Mr. Poole heard patiently an account of our dilemma, and then quietly remarked: "You must grind without pressure; come to Wilmington, and I will show you how to do this." He prepared wheels suitable for grinding the points of calipers, a thing he had never done before for any one, and extended many courtesies and kind acts that will never be forgotten.

The problem of making fixed calipers was solved, and gage-making rendered possible; first by Prof. John E. Sweet, who discovered and pointed out the nature of the difficulty in grinding the contact points, and by J. Morton Poole, who removed this difficulty by his knowledge of abrasive processes.

As soon as accurate calipers could be ground, orders were taken and the business began, one of the first orders being for a railway works in England, and the second for the Baldwin Locomotive Works, at Philadelphia. It was sixteen years from the time of making the first set of calipers at Columbus, O., and $18,000 had been spent in implements, experiments and other expenses.

[After all his work to found the American Standard Gage & Tool Works, Richards sold out to the Betts Machine Co., Wilmington, DE, in 1879. Betts moved the business to Wilmington and sold it to the Taylor-Rice Co. in 1895. Taylor-Rice sold out to the John M. Rogers Boat, Gauge & Drill works about 1897. Rogers continued the line until sometime after World War I.]

APPENDIX II

DARLING & SCHWARTZ

The following letters appeared in the *American Machinist* magazine in March and April, 1914, in response to a letter requesting information on a scale marked "D.& S. BANGOR, Me." Comments in brackets have been added by the author.

I noticed an item on page 252, referring to an 18-in. scale, marked "D.& S., Bangor, Me.," and asking for information as to who made it.

This scale was no doubt made by Darling & Schwartz, who at the time referred to were manufacturing scales, squares and the like in Bangor. Samuel Darling was the mechanical member of the firm, and Mr. Schwartz furnished the money. Mr. Darling built a graduating machine in 1852, and this was in use in Bangor until 1868. As to the accuracy of work done on this machine the following letter will testify:

U.S. Navy Yard
Washington, April 10, 1855

This is to certify that I have seen one of Darling & Schwartz's Test Scales compared with a Government Standard Scale. A card accompanying the latter stated the first foot to be correct, the second foot to err the 0.0002 of an inch, the third foot to err 0.0007 of an inch, and on comparison with Darling & Schwartz's scale, an error was detected, perhaps about as much as above stated, thus showing great correctness in the scales made by D.& sincerely.

Darling & Schwartz's scale agrees exactly with foot No. 1, which is stated to be correct.

H. Hunt
Chief Engineer U.S.N.

Previous to the building of Mr. Darling's graduating machine, Joseph R. Brown, of the firm of J.R. Brown & Sharpe, designed an automatic graduating machine, which was put into use on Aug. 22, 1850. As far as is known, this was the first automatic graduating machine in America. Besides being used for graduating scales, this machine was used to graduate the first Vernier calipers, which were put on the market in 1853. It is still in successful operation.

In 1866 Mr. Darling formed a partnership with J.R. Brown & Sharpe, for the manufacture of machinist's tools, and in 1868, Mr. Darling's plant (together with the graduating machine) was moved to the Brown & Sharpe works, in Providence, R.I. This machine is still running, and is being operated by John E. Hall, the same man who has run it continuously ever since—a period of more than 50 years.

Mr. Darling early made a reputation for high-grade work, and was awarded numerous medals for exhibits of his products. Among these was an award from the Franklin Institute in 1856, which is inscribed: "To Darling & Schwartz, Bangor, Me., for Machinists' Tools, 1856." J.R. Brown & Sharpe were also awarded a medal at this same exposition. In 1862 a medal was awarded Darling & Schwartz for an exhibit shown in London.

L.D. Burlingame
Providence, R.I.

[Luther Burlingame was Chief Draftsman for Brown & Sharpe.]

The scale marked "D.& S. Bangor, Me.," regarding which the W.P. Davis Machine Co., of Rochester, N.Y., made inquiry on page 252, was made by Darling & Schartz, who formerly carried on the business of making accurate hard-edge try-squares, steel scales, steel straight-edges and the like in Bangor, Maine. Mr. Darling, who was a fine mechanic, was a pioneer in this kind of business. Mr. Schwartz, the financial partner, carried on a drug store. Afterwards, Brown & Sharpe bought out Mr. Schwartz's interest and moved the business to Providence, R.I., and after Mr. Darling's death, purchased his widow's interest also.

Making Hard-Edge Try-Squares

A friend of mine who worked for Mr. Darling before the business was removed to Providence, has furnished me with some facts about it that may be of interest. One of the lines of tools Mr. Darling made was accurate hard-edge try-squares. The blades or tongues were first heated and then quickly clamped between chills which touched only at the edges, the center being relieved; this hardened the edges of the tongues and left the center soft. The blades were then ground. The grinder consisted of a commmon grindstone, some 6 ft. in diameter, under which a platen traveled; the grindstone was plentifully supplied with water meanwhile. [This machine is believed to be the first surface grinder and was patented by Darling on August 30, 1853.] The handles, or stocks of these squares, were made of two pieces of white iron castings (unannealed malleable iron—"glass hard") between which was fastened a piece of steel about .001 in. Th icker than the tongue (this was to allow room for the solder, as will be explained later on) and nearly the width of the tongue shorter than the castings, making a slot into which the blade was to be inserted, and fastened by sweat-soldering. [This process was patented by Darling, October 6, 1857.] In grinding these white iron castings, it was found necessary to keep fine sharp sand sprinkled on the grindstone, in addition to the water, to prevent glazing of the stone.

After being ground true, one end of a blade and the slotted end of a stock were immersed in a pot of melted solder. After becoming suffciently heated, the tongue was slipped into the slot and was slipped back and forth until both the tongue and the slot were thoroughly tinned, the proper flux being employed meanwhile. Then both were quickly withdrawn from the solder pot, and were accurately squared by being quickly held against three hardened pins, properly located. No other fastening was necessary, one could grasp a square by the blade and drive nails with the end of the stock without starting the solder, when the work was properly done. After everything else was finished, the panels that were cast on the outside of the stock were filled with some hard wood, such as lignum-vitae or rosewood.

I have in my possession one of these 12-in. squares, which has been in continous use for over 30 years and it scarcely shows any signs of wear. At one time, a 12-in. square like this retailed for $8. [A day's pay for a machinist at that time was about $1.25-1.50.]

A Test of Straight-Edges

When Mr. Brown came up to Bangor to look over the place, Mr. Darling called his attention to a pair of straight-edges which were some 5 or 6 ft. long, 5 or 6 in. wide and $\frac{3}{8}$ in. thick, made of bar steel and ground true. They were set up on edge, one on top of the other, and Mr. Brown, with a magnifying glass, was trying to find a place where he could see the light through between then, but admitted that he was unable to do so. After a few minutes' conversation, Mr. Darling asked him to take another look, and to Mr. Brown's suprise, he could see the light through nearly the whole length. Mr. Darling showed him that he had merely held his hand on top of the upper straight-edge and the warmth of his hand had expanded the top edge sufficiently to temporarily cause this variation from accuracy. [This story is probably apocryphal. Steel, of course, expands with temperature increase, but the amount of heat and the length of time required to make a measurable difference in a piece with the mass described would be well in excess of the above description.]

At one time Mr. Darling had a large lot of lead-pencil sharpeners to make, and the bevel of the little knives at the cutting edge was made on this grinder. A special platen was made, having a number of narrow slanting grooves, running crossways, which were filled with the unsharpened knives. They were merely dropped in, not fastened at all, then they were all nicely beveled and sharpened by sliding the loaded platen under this revolving 6-ft. grindstone. Small work for a big machine!

W.A. Sylvester
Reading, Mass.

Another letter, this one from May, 1884, adds a bit more to our knowledge of Darling & Schwartz.

From C.E.S., Boston, Mass.—I have in my notebook an account of how a lead screw was corrected some 30 years ago, and as the particulars were obtained from one who at that time was an apprentice at the works where the operation was performed, the orginal screw, moreover, being still in use, it may prove of some interest. In the year 1854 Mr. H. Sibley, who was carrying on the machinist business, received an order to correct a screw for a Maine firm engaged in the manufacture of graduated rules and squares for machinists' use. [This could only have been Darling & Schwartz where Darling was building his second graduating machine in 1854.] The lathe upon which the screw was cut was made, I believe, in 1850, by S.C. Coombs, of Worcester, Mass. The lead screw of this lathe has five threads to the inch, square thread, of $15/16$-inch outside diameter and threaded for a distance of 45 inches. The original screw was found to be short of what it would have been if it had been correct in length. Each inch was slightly short, and in a total length of 3 feet it was found to be $1/32$ inch short. The spaces between the threads of the screw at the bottom were stretched by paning [peening]. This paning extends almost in every thread of the entire length of the screw, and the time expended in doing the work was about 50 hours.

[For those of us who are hazy about the use of the peen side of ball or straight peen hammers, this reminds us that peening was an important part of the machinist's art, often practiced in the 19th century. It was commonly used to straighten machine parts and to correct shape or size when errors were found. It's easy to see why peening is a lost art. Expending 50 hours of labor to correct a lead screw would now be considered the height of folly.]

After this screw was corrected, Mr. Sibley cut two solid nuts about 8 inches long each. These were connected so as to form one continuous nut, which was then used upon the screw, corrected, and employed to cut a second screw and several others. At present the third screw thus cut is saved and used only as occasion requires. The screw cut for the Maine firm, now the Brown & Sharpe Manufacturing Company, of Providence, R.I., was accepted and gave entire satisfaction.

APPENDIX III

JOHN COFFIN OBITUARY

The following obituary, written by John E. Sweet, founder of the Straight Line Engine Co. and one of the foremost mechanical engineers of his time, appeared in *American Machinist* magazine of September 19, 1889. It is of astounding length for a 33 year old, indicating the esteem in which he was held in the mechanical engineering field.

John Coffin, whose communications have appeared in the *American Machinist* from time to time during the last few years, died at Johnstown, Pa., on the 3d of September, at nearly 33 years of age. Born at Chatham, New York, in 1856, he secured a common school education, and the rudiments of the machinists' trade, and went to the Indian Territory with his father, where they were engaged in moving the Pawnee Indians. While there the thought came to young Coffin that he had not all the education that might be useful to him, so returning, he, after making some preliminary preparations, entered Cornell University. Here, although showing marked mechanical genius and wonderful originality in his geometrical demonstrations, he was dropped from the University roll by the authorities, at the end of the first year, admittedly because of the unfashionableness of his spelling. Leaving the University, he found employment in different machine shops in Ithaca and Syracuse. and in 1880 took the position of foreman in the works of the Straight Line Engine Co

While in the Indian Territory, and the various machine shops, his genius and mechanical skill helped himself and others out of many a difficulty, and while at the engine works he invented his averaging instrument, now well known to engineers, and designed the throttle valve used on all Straight Line engines. Later he invented a graduating and other machines used in the manufacture of machinists' scales, by the use of which thousands of these scales have been made and are now in the hands of machinists, bearing the stamp of Coffin & Leighton

Some time in 1881 he entered the service of the Cambria Iron Works, Johnstown, Pa., where he held the positions of draftsman, foreman, and, later, investigator and inventor. The work which he accomplished there that was the most valuable and will give him the most lasting name, was that of treating railway axles and rails, by a peculiar way of heating and cooling, which increases their ability tenfold to resist the blows of the drop. The human lives and property this one discovery is destined to save ought to make his name immortal. Another of his practical inventions—that of a new method of annealing and galvanizing fence wire—had just been put in opertion at the works, when the great disaster [the Johnstown flood] blotted it out.

Large works like the Cambria give scope to the inventive genius of any man, however gifted, but, in addition to this, Coffin's work was blended with scientific investigation. Many of his discoveries in the behavior of steel came out of his experiments in the treatment and tempering of the steel hoops for the government cannon, and among other discoveries was that which determined pretty conclusively what takes place when a piece of steel is hardened and when annealed; also that two pieces of clean steel, perfectly fitted and heated to a red heat while in contact, will weld together. Another discovery was that, when a piece of steel is highly heated and allowed to cool, that, at at certain stage of the cooling, a change takes place, and the piece becomes momentarily hotter and softer than before, and more pliable than at a time when at a much higher temperature.

JOHN COFFIN OBITUARY

Coffin was married about a year ago to Miss Elizabeth H. Fussell, of Radnor, Pa., and they were living in a new and completely happy home at Moxham, where they missed the immediate result of the great flood, but which led to a result that brought on fever, and the death of one of the brightest and best of men I ever knew.

APPENDIX IV

SENSITIVENESS OF TOUCH

A still ongoing argument among machinists is the validity, or lack of validity, of the machinist's "feel" or ability to detect very small variations in dimension or form using only the fingers. There is a related "feel" by which machinists can detect very small deflections in measuring instruments when in use.

I have seen individual cases of this "feel," which many say does not exist, both proven and disproven by informal, on the spot, tests.

The following article, from the *American Machinist* magazine of March 7, 1907, relating a story from 1892, shows that the argument has been going on for over 100 years.

During the week of the dedications of the exposition buildings at Chicago, Ambrose Webster and a well known manufacturer of steel balls were there, and were met one evening a the hotel by the president of the American Machinist Company. The subject of ball bearings was introduced, and the ball manufacturer, taking a hardened and ground ball from his pocket, handed it to Mr. Webster, who, rolling between his fingers, remarked: "Why don't you make them round?" "It is round," was the reply, to which Mr. Webster answered, "No, it is not round—I can feel that it is out of round." The manufacturer was unable to detect any want of truth, and the result was a trifling wager between him and Mr. Webster, who said that if allowed to take the ball home he would measure it, and determine whether or not it was true, and how much it was out, if at all.

A few days later Mr. Webster called at this office, armed with a fine micrometer caliper, made by A.J. Wilkinson, of Boston, and said: "I am now prepared to win the wager made in Chicago, as I can demonstrate that the ball is not round. It has around it what may be termed an equatorial line which stands above the surface of the ball, or, in other words, there is a belt extending around it, the diameter of which is greater than that of the ball when measured at any other point than through this belt; and I can pick it out every time by the sense of touch, and place it in the caliper so that it will not pass through, though the caliper be set so that the ball will easily pass through when turned so that this belt does not come into contact with the screw."

This Mr. Webster was successful in doing at every attempt; and Fred J. Miller, then associate editor of the *American Machinist*, who was called in to see what he regarded as a remarkably delicate sense of touch, found after a few trials that he could himself easily detect the belt by rolling the ball in his fingers, and demonstrated that he did so by placing it correctly in the calipers at every trial. Mr. Webster, again taking the ball and rolling it between his fingers, said there was one feature about it that he had not before noticed, i.e., there seemed to be at a certain point in the belt a slight projection, or, as he described it, a mountain; and after a little examination both he and Mr. Miller demonstrated that they could define the location of this mountain by the sense of touch, and theyboth placed the ball in the caliper repeatedly so that it just touched, when, if turned very slightly from that position in any direction, it could pass through without touching.

A bench measuring machine in Mr. Webster's shop at Waltham (The American Watch Tool Company), showed that the ball, which was $\frac{3}{8}$ inch nominal diameter, actually measured through its smallest diameter 0.37512 inch; through the mean or diameter of the belt, 0.3752 inch; and through the mountain and belt, or its largest diameter, 0.37539 inch. Thus it is seen that the ordinary diameter through the belt is 0.00008 inch greater than through the smallest part, and that the mountain is 0.00019 inch above the belt, the diameter measured through the mountain being 0.00027 inch greater than the smallest diameter.

The tests showed, we think conclusively, that variation in size or shape amounting to less than 0.0001 inch can be detected by the fingers, where, either naturally or by training, the sense of touch is highly developed. So far as the matter has been investigated, it does not seem that such sensitiveness is common, few persons being able to detect any variation in sphericity, or, where they think they do detect it, they fail to prove it by placing the ball in the caliper correctly.

Shortly after the account of the tests as noted above was brought to the attention of Professor Sweet he wrote us as follows:

> "Having some friends engaged in the hardened-steel ball business, and being interested in machine shop measurements, I was naturally interested in your account of Ambrose Webster's thumb and finger caliper, and could only reconcile your statement to my conception of the possibilities by fancying you had one too many ciphers on the account. It takes a fairly good man to detect 0.0001 inch difference in the size of a piece with a common calipers, and to concede that a man could do better with his fingers, was, at first, straining to one's belief.
>
> Picking up some sample balls at hand, we noted, after but a monment's fingering, that they were not round, and one appeared to have the same peculiarity noted in your account. While we had no means at hand for determining the absolute variation, it was easy with the small measuring machine to determine that the variation was not far from 0.0001 inch."

APPENDIX V

MICROMETER CALIPERS OF 1917

The illustrations included in this appendix are from an article by F. Server which appeared in the November, 1917, editon of *Machinery* magazine. Note that the 1917 machinist had essentially the same selection of micrometers, with the exception of a digital readout feature, as the machinist of 1997.

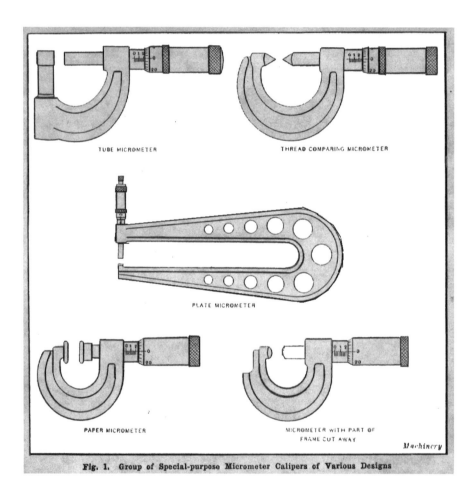

Fig. 1. Group of Special-purpose Micrometer Calipers of Various Designs

MICROMETER CALIPERS OF 1917

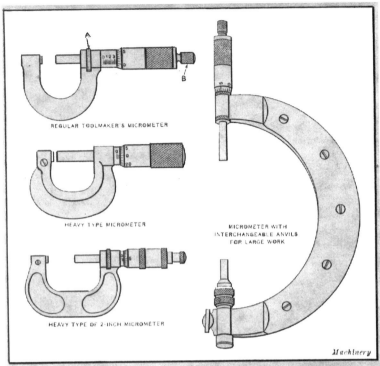

Fig. 2. Group of Standard Micrometers of Different Types and Sizes

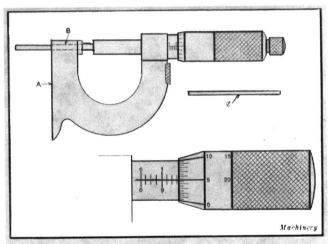

Fig. 4. Micrometer provided with Double Set of Graduations and Depth Plug

MICROMETER CALIPERS OF 1917

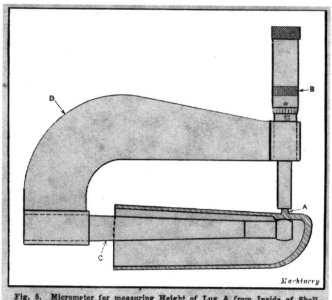

Fig. 5. Micrometer for measuring Height of Lug A from Inside of Shell

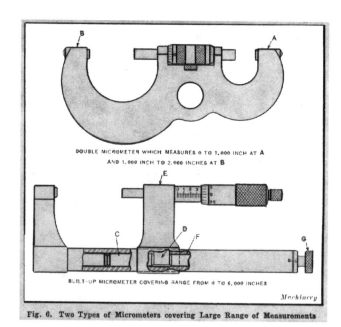

Fig. 6. Two Types of Micrometers covering Large Range of Measurements

MICROMETER CALIPERS OF 1917

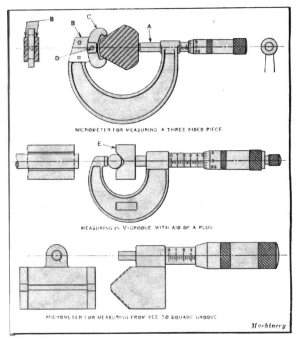

Fig. 7. Three Special Applications of Micrometer Calipers

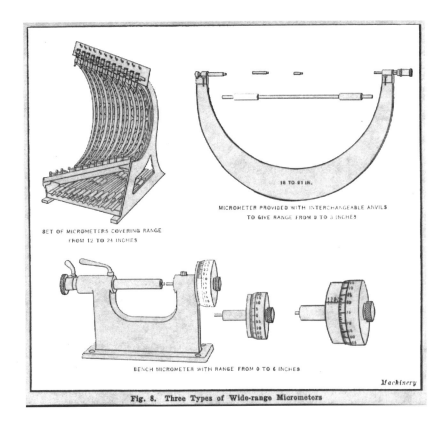

Fig. 8. Three Types of Wide-range Micrometers

MICROMETER CALIPERS OF 1917

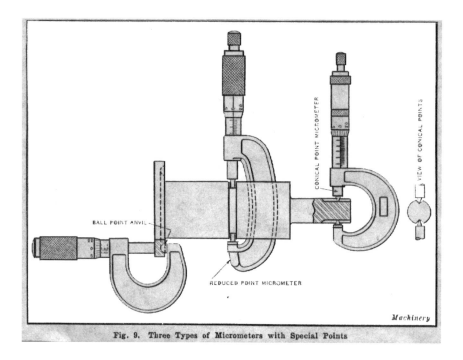

Fig. 9. Three Types of Micrometers with Special Points

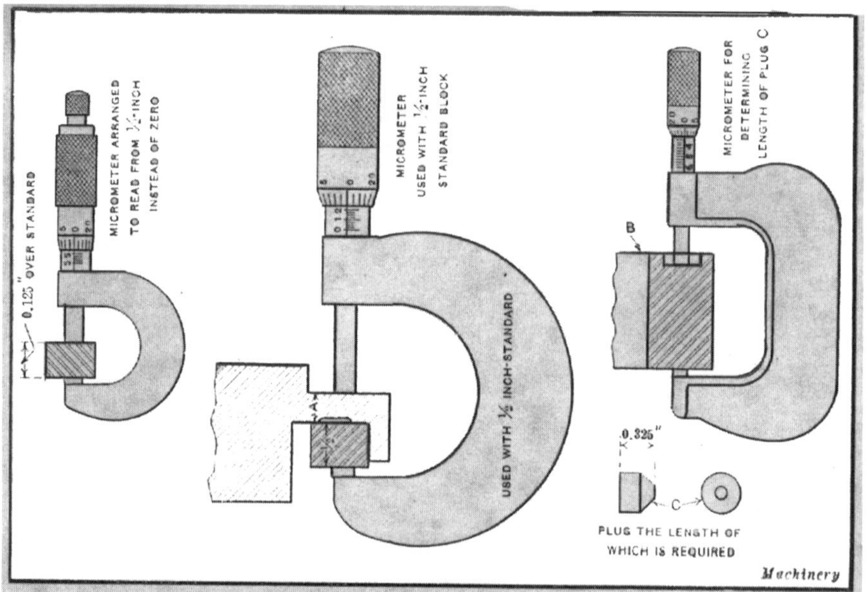

Fig. 10. Micrometers that take Readings from Some Other Point than Zero

MICROMETER CALIPERS OF 1917

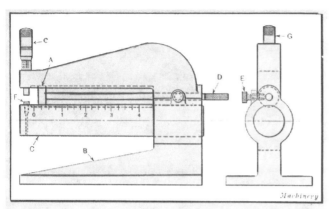

Fig. 11. Micrometer for measuring Thickness of Tubing at Intervals of One Inch

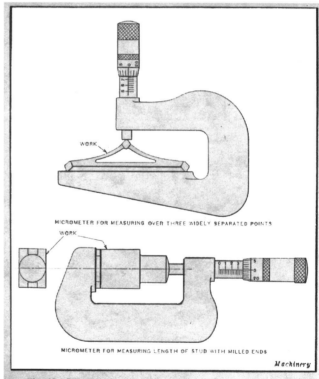

Fig. 12. Micrometers with Wide Anvils to afford Support for Work

MICROMETER CALIPERS OF 1917

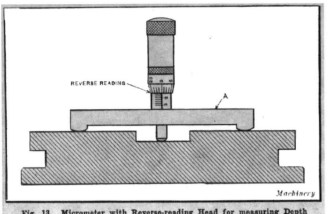

Fig. 13. Micrometer with Reverse-reading Head for measuring Depth of Slot

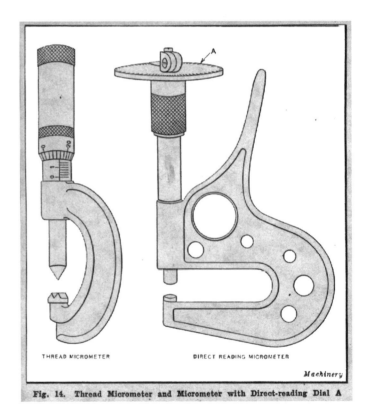

Fig. 14. Thread Micrometer and Micrometer with Direct-reading Dial A

MICROMETER CALIPERS OF 1917

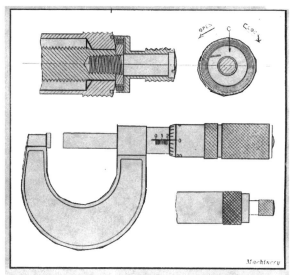

Fig. 15. How Friction Stop is applied to Micrometer Thimble

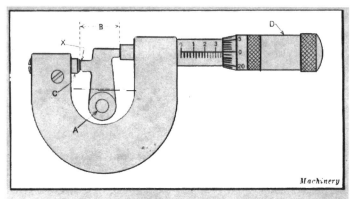

Fig. 16. Micrometer for obtaining Dimension B of Small Lever

APPENDIX VI

MEMORIES OF AN ALL-ROUND MACHINIST
By John J. Grant

John J. Grant (1844-1934) was one of the foremost machinists and machine tool designers of his time. In addition to working for several machine tool companies, he founded Grant & Bogert, Grant Machine Tool Works, Grant Anti-Friction Ball Co., Grant Machine Tool Co., Grant Tool Co., Grant-Lees Co, Grant Automatic Machine Co., and Grant & Wood Mfg. Co. His description of a machine shop of the 1860s, published in the *American Machinist* magazine July 11, 1912, and October 30, 1923, gives us a clear picture of the machines and methods of the day. Material in brackets has been added by Ken Cope.

After leaving Brown & Sharpe's shop [in 1859] I went into a woolen mill, and helped the all-round man who did the repairs. I had the starch taken out of me when I encountered a job that seemed impossible to do with the tools we had. It was done, however, and I had sense enough to see that if I was ever to become the all-round man I hoped to be I must be able to do these jobs and others that would perhaps put me in a tight place.

I went to Northampton, Mass., and hired to a man who ran a foundry, machine shop, sash and blind shop—Old Bill Clapp. Nothing got away from him, he would take anything to make or repair in that shop from cleaning a watch to building an 18x42 steam engine. I worked there for three years and laid the foundation for all the success in mechanics that has come to me.

The shop had a fair equipment of tools, as shops were supplied in those days. As far as I can remember there were three small engines lathes of 14 to 16-in. swing, wooden beds with the iron V-ways and a chain feed; a 16-in. lathe a little more modern, which I remember had a curious arrangement of the speed-reducing gears, all inside the one cone pulley, and one 16-in. lathe made by the Hadley Falls Machine Co., of Holyoke, Mass. This was our star lathe and used for the best work. We had several other lathes all of the same general design, but only one with the feed in the apron.

A 24 in. x 20 ft. bed lathe was used for a shafting lathe and I have turned miles of shafting 2 to 3 in. in diameter on it, the rate of about 30 ft. per day. We used all iron shafting in those days and when Old Bill had taken a job extremely low, he bought correspondingly bad iron for the job. This iron was full of sand streaks, so we were obliged to stop the lathe every foot or two, and chip out the sand streaks to prevent the tools wearing too fast.

We turned all shafting through a ring not cut apart, and it required constant watching to keep the size close enough to fill the ring, and free enough to pass through. The diameter of the follower ring was enough larger than the finished size of the shaft to allow filing. We had standard rings bored out to inside calipers set to a scale with the corners worn off, but it was a standard even though it did not match the other fellow's who made shafting. Old Bill used to chuckle when we told him about it, and say: "Well, that's all right; the people who buy this shafting will have to come to me for pulleys."

We had a 8-ft. lathe built on a stone foundation having the plate of cast iron leveled with hot sulpher. These ways were not so accurate but we did some good work, as the lathe was strongly geared.

The spindle was of cast iron about 8 in. in diameter and 10 in. long in the front box, which was also of cast-iron. After many years use it had a beautiful surface. The faceplate was keyed to the spindle. While doing a job I found that the faceplate was loose, so I took off the front spindle cap, backed out the key and shimmed under it with a piece of tin. This threw it out of true nearly a quarter of an inch and we had to use it in that shape for at least one year before I could get permission to true it.

Our drill press was home-made, being constructed from the headstock of an old wood-turning lathe. There was no quick return to the spindle. The quill was threaded on the outside and ran through the front box. It had a handwheel at the top for feeding. The spindle was driven by a pair of cast bevel gears at the top. One of these was mounted on a horizontal shaft which went through a hole bored through the post to which the drill press was bolted. There was a cone pulley on this horizontal shaft and another shaft on the bottom of the post with a cone and tight and loose pulleys.

The table was a large angle casting bolted to the post and made so that it could be moved up and down. The spindle had a straight hole and large drills or smaller collets were fastened with a setscrew. We had no twist drills, reamers, tap, dies, or milling cutters expect home-made ones, and some of these were fearfully and wonderfully made.

We had a suspension drill that was one of the handiest tools I ever saw. A tool of this description would be a valuable tool for all shops doing large work in the present day.

We had one planer 24x24in.x6ft. made by Gage, Warner & Whitney of Nashua, N.H. The architectural design was rather elaborate, but the machine designer was certainly abroad when the drawings were made (if there ever were any). We had more trouble to the square inch to make that planer behave than any tool I ever saw. The platen was continually running off the rack, and tipping endwise to the floor.

The elevating screw was in the center of the crossrail and after each raising it was necessary to set the crossrail parallel with the platen, with a surface gage. Of course, we changed it as seldom as possible, using long tools for the work.

There was one tool in the shop that stood out from all the others and had nearly all the refinements of tools of its class made today. This was a 42x42in.x10ft. planer made by Slate & Brown of Windsor Locks, Conn. It was designed by Dwight Slate, one of the foremost mechanics of his day. I have often talked to him about this tool and it pleased him to have me enumerate its good points. It was the first friction feed box, afterward applied to Pratt & Whitney planers.

We had a gear-cutting machine bought second-hand from the old Lowell Machine Shop. It was built on a wooden frame, had an index plate about 30 in. in diameter, and had been used so long that several rows of holes looked more like a groove running around than drilled holes.

Our gear cutters were all straight face, commencing with about 30 deg. included angle, and guessed at for the less number of teeth. We filed most of the teeth to shape, often putting them on studs on a plate and running them together, getting the shape that way. The only man who knew a thing about gearing was the pattern maker who used to shape the teeth by eye.

Our other tools were a bolt cutter, hand lathe, and, of course, the usual grindstone under the stairs. At first it was difficult for me to grind a tool on account of the stone running out of true. You simply had to hold the tool solid on the rest and let the high places on the stone hit it when it came around, which by the way is often the case in the shops of today. The only difference is that the speed of the modern grinding wheel hits it more rapidly.

We had a first-class blacksmith and tool dresser. He was a godsend to any shop in those days, when material cost so much and lathes and other tools were not made strong enough to remove a heavy cut. After a few months in Old Bill's shop, the Civil War was declared and from that time on for the next three years, we were kept hustling. We did a large amount of work for the government such as gun-barrel drilling machines, barrel-reaming machines, lapping and polishing machines, rolls for forging bayonets. Everything was on the jump, and we certainly averaged 14 hours per day during that time.

We had no addition to our equipment except such small tools as were actually necessary to produce the results as called for by government inspectors. Old Bill fought the making of every one, as he used to say: "You machinists ought to do anything, with all the tools we have in this shop." Most of the machines we made were well built and I would not be suprised to find many of them still in use at the Armory at Springfield, Mass. [No doubt they were, 50 years was well within the normal life span for machine tools of the time.] In that shop we did such an endless variety of work that it was no uncommon thing to see a set of drivers from one of the Hinckley & Drury locomotives used on the Connecticut River R.R. in the big lathe, and a man close by repairing a small sewing machine. We had no specialists, every man was supposed to know how to handle every machine in the shop. No one man had a machine of his own but took any that was free at the time his job was given to him.

When I had finished my three years post graduate job in Old Bill Clapp's shop, a friend had a place for me as tool maker in the Florence Sewing Machine Co.'s shop. This company was the outgrowth of an old shop formerly making turbine water-wheels and other machinery. The manager who was there when I arrived was certainly in a class by himself, as he had no more idea of what was wanted in the matter of tools to manufacture sewing machines, than a fifteen year old boy. One of his hobbies was to come around and measure with a two-foot ivory rule, graduated to $\frac{1}{16}$ in., some fine piece of work on which we had used a vernier caliper and say, "Well that will do but we must improve on it."

Our machine tools were mostly ones that had been sent in from the manufacturing department to be rebuilt but we managed to keep them. There were no universal milling machines or grinding machines, and small tools were all homemade. The first good lathes we had were two 13 in. and two 16 in. with Dwight Slate's taper turning attachments. They were the first lathes sent out from the Pratt & Whitney shop and were at that date the acme of perfection in design and workmanship. I had the honor of setting them up and a short time afterward had the pleasure of meeting both Mr. Pratt and Mr. Whitney. The acquaintance with these men changed the whole course of my life and there was never a break in our friendship.

I often look at the fine tools in the windows of the first class hardware stores and wonder how we succeeded in doing accurate and difficult work with the makeshifts we had. I am not exaggerating when I say it would be a wonder if out of one hundred good machine shops of that day, you could find two 1-in. plug gages that would measure alike, although called standards. The vernier caliper was at that time the universal standard and there were as many differences in any particular size as there were men that set the vernier. The micrometer changed the whole of that, as the general mechanic can set it to one ten-thousandth of an inch.

Those were stirring times and at the closing of the Civil War, attention was given to the development and manufacture of many new and accurate labor saving small tools, most of them emanating from the gun and pistol factories and the men who had been there employed.

Morse, of twist drill fame, with whom I worked at the Florence shops, had designed machines to manufacture twist drills. After he had gone to New Bedford, Mass., and started a small shop for making the drills, he came to the Florence shop and brought drills of the sizes used there, for trial. It would be laughable today to hear the comments of the tool makers on the drills, and the probable failure of the business of making and selling store drills. After testing the drills for several days, Morse went back with a good-sized order and no more drills, except special ones, were made in that shop.

Soon after that Ed. Beach brought out the Beach drill chuck which was a marvel of accuracy and has never been excelled in workmanship.

The late James M. Carpenter, and my humble self left the Florence shop on the same day (I think it was the first day of April, 1864) without either knowing what the other intended to do, but with the same purpose in view. It has been a question ever since who put the first taps and dies on the market. I started a shop in Northhampton, Mass., and Carpenter started one in Pawtucket, both of us making taps and dies. A short time after, I moved my business to Greenfield, Mass., and invented the Grant screw cutting die and the Lightning screw plate. Shortly after that I organized the Wiley & Russell Mfg. Co. which was the nucleus of what is now the Greenfield Tap & Die Corporation.

APPENDIX VII

1915 UNION CALIPER CO. CATALOG

This is one of the few catalogs issued by the Union Caliper Co. during its short, eight year life. It is also the last, issued just before the company reorganized as the Union Tool Co. in 1916.

(continued on 15 following pages)

IN PRESENTING this catalogue, we wish to call your attention to the several new lines of tools shown. In addition to our well-known line of goods shown in our previous catalogues, will be found Tempered Steel Rules, Combination Squares, Hack Saw Frames, Key Seat Rule Blocks, Thickness Gauges, our new line of Quick Adjusting Calipers and our new Adjustable Boring Tools.

In designing these new tools, it has been our aim to have them superior to anything yet shown by any manufacturer, keeping in mind all the time our business policy to FURNISH HIGHEST POSSIBLE QUALITY AT REASONABLE PRICES.

We take this opportunity of thanking our many customers for their co-operation, and ask for future business on the basis of the QUALITY OF GOODS and the SERVICE WE CAN RENDER.

We are not trading, however, on what we have done or what we are going to do, but what we CAN do for you at the present time, by furnishing RELIABLE TOOLS at MODERATE PRICES.

Our tools are carried in stock by nearly all the best tool dealers, and can generally be procured from them most advantageously. If, however, your dealer will not supply you with our tools, write us, and we will see your wants cared for.

If, when ordering, you give us no shipping instructions, we shall use our best judgment in routing your shipments.

Dealers with no commercial rating, must furnish satisfactory references or cash with order.

WE GUARANTEE EVERY TOOL WE MANUFACTURE TO GIVE SATISFACTION FOR THE PURPOSE WHICH IT IS INTENDED. NO IF'S OR AND'S ABOUT IT.

UNION CALIPER COMPANY

ORANGE, : MASSACHUSETTS

MANUFACTURERS OF

CALIPERS, DIVIDERS, TAP WRENCHES, NAIL SETS, CENTER PUNCHES, TEMPERED STEEL RULES, COMBINATION SQUARES, HACK SAW FRAMES, KEY SEAT RULE BLOCKS, THREAD GAUGES, THICKNESS GAUGES.

Complete Line of Tool Holders for Turning, Planing, Boring, Shaping, Slotting, Cutting-Off, Side Cutting, Threading, Key Seating, Lathe Dogs, Drill Holders, Machine Vises and Screw Machine Products.

EVERY TOOL of a SUPERIOR QUALITY and GUARANTEED to give SATISFACTION

UNION ADJUSTABLE TAP WRENCH
No. 25

A handy quick adjusting wrench for taps, drills, reamers and tools of a like nature. It is provided with a countersunk center for use in a lathe. Made of tool steel properly tempered.

Two Sizes

PRICES, No. 25

No. 1.	Extreme Capacity 1-4 inch taps	$ .50
No. 2.	" 1-2 " "	1.00

Packed one in a box.

THE JAWS HAVE JUST THE RIGHT SPRING TO OPEN AND CLOSE PROPERLY

UNION SPRING CALIPERS AND DIVIDERS

THESE HAVE MANY POINTS OF SUPERIORITY

THE SPRINGS are constructed to fit the hand and prevent side deflection in the legs of the tool, and are of ample strength to retain the position in which the tool is set. The jar of the machine or rough handling will not change the size they are set at.

THE NUTS are hardened and run smoothly.

THE SCREWS have flat top threads for strength and wearing qualities and also prevent the collection of dirt.

THE CALIPER POINTS are of correct shape and proper size. The Divider points are especially tempered for wearing and cutting qualities.

OUR GUARANTEE MEANS SOMETHING

1915 UNION CALIPER CO. CATALOG

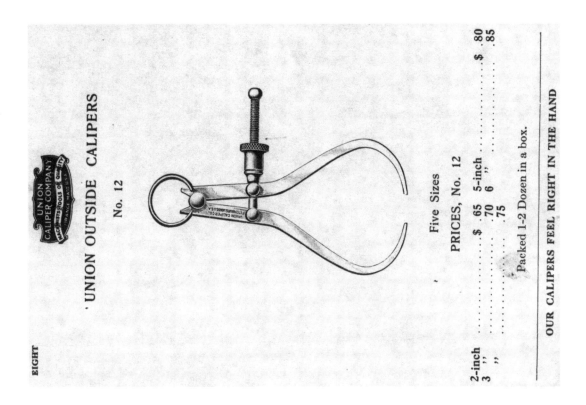

UNION OUTSIDE CALIPERS
No. 12

Five Sizes
PRICES, No. 12

2-inch	$ .65	5-inch .... $ .80
3 "	.70	6 " .... .85
4 "	.75	

Packed 1-2 Dozen in a box.

OUR CALIPERS FEEL RIGHT IN THE HAND

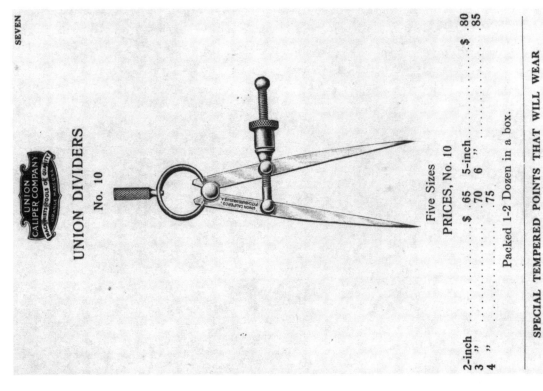

UNION DIVIDERS
No. 10

Five Sizes
PRICES, No. 10

2-inch	$ .65	5-inch .... $ .80
3 "	.70	6 " .... .85
4 "	.75	

Packed 1-2 Dozen in a box.

SPECIAL TEMPERED POINTS THAT WILL WEAR

UNION FIRM-JOINT CALIPERS

There are many superior features in this line, including our new adjustable triple friction joint; easily adjusted with a common screw driver to any desired tension.

The contact points are sufficiently hard to make them very sensitive to the touch when calipering.

The weight, shape and proportion are correct.

These, like all of our other tools, are made to meet the approval of the most exacting mechanics.

REMEMBER ALL OUR TOOLS ARE GUARANTEED

UNION INSIDE CALIPERS
No. 14

Five Sizes

PRICES, No. 14

2-inch	$ .65	5-inch	$ .80
3 "	.70	6 "	.85
4 "	.75		

Packed 1-2 Dozen in a box.

OUR CALIPERS FEEL RIGHT ON THE WORK

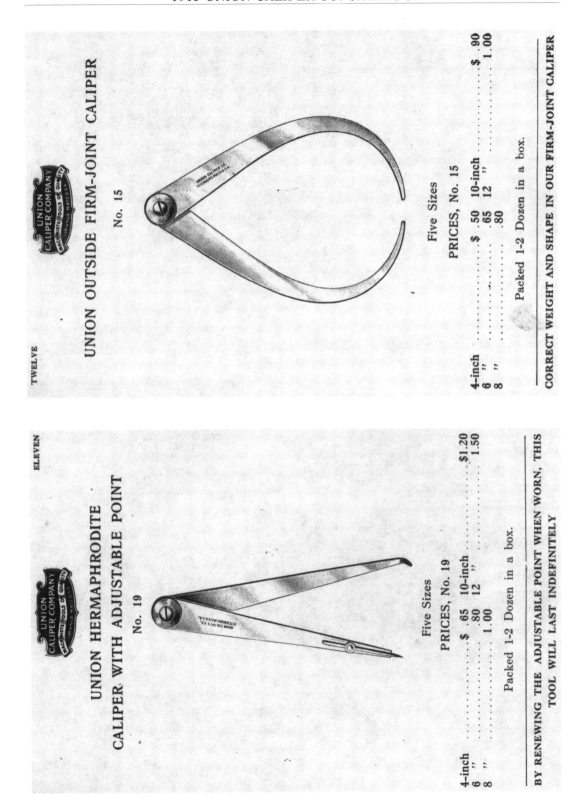

UNION OUTSIDE FIRM-JOINT CALIPER
No. 15

Five Sizes

PRICES, No. 15

4-inch		
6 "	$ .50	10-inch .... $ .90
8 "	.65	12 " 1.00
	.80	

Packed 1-2 Dozen in a box.

CORRECT WEIGHT AND SHAPE IN OUR FIRM-JOINT CALIPER

TWELVE

UNION HERMAPHRODITE CALIPER WITH ADJUSTABLE POINT
No. 19

Five Sizes

PRICES, No. 19

4-inch	$ .65	10-inch $1.20
6 "	.80	12 " 1.50
8 "	1.00	

Packed 1-2 Dozen in a box.

BY RENEWING THE ADJUSTABLE POINT WHEN WORN, THIS TOOL WILL LAST INDEFINITELY

ELEVEN

UNION INSIDE FIRM-JOINT CALIPER
No. 17

Five Sizes

PRICES, No. 17

4-inch		$ .50
6 "		.65
8 "		.80
10-inch		$ .90
12 "		1.00

Packed 1-2 Dozen in a box.

OUR TRIPLE FRICTION JOINTS STAY WHEN SET

UNION QUICK ADJUSTING SPRING CALIPERS AND DIVIDERS—ELLIS PATENT

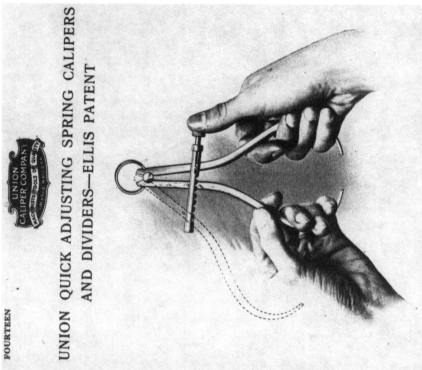

The quickest and best caliper adjustment that has ever been offered to the mechanics. Instantly adjusted from maximum to minimum, or any part of the calipers' capacity, by simply raising the adjusting bar, disengaging the bar from the stop pin, as shown in the above illustration.

SPEED AND ACCURACY

UNION QUICK ADJUSTING INSIDE SPRING CALIPERS

Have Positive Transfer Features

No. 514

Made in Four Sizes

PRICES

3-inch Each		$1.15
4 " "		1.35
5 " "		1.55
6 " "		1.75

OUR GUARANTEE MEANS THAT YOU MUST BE SATISFIED

A DESCRIPTION OF THE UNION QUICK ADJUSTING CALIPERS AND DIVIDERS

A brief description of this caliper will be of interest to all mechanics.

The adjusting bar or quick adjustment on this line of calipers and dividers, is formed from a solid piece of round stock. A slot milled out of the center, in which the caliper legs are fitted just tight enough to give them a free movement and support, eliminating side motion in the caliper legs, making them very stiff and strong, without being heavy and cumbersome.

There are several notches cut in this adjusting bar, that engage a stop pin in one of the caliper legs. By simply raising this bar, disengaging it from the stop pin, calipers may be opened or closed to approximate size desired.

Fine adjustment is then obtained by a fine thread screw of ample diameter to insure its wearing qualities, giving the advantage of a very fine, sensitive adjustment, much finer than is obtained on other types of calipers and dividers. This screw needs only $\frac{1}{4}$ inch to $\frac{3}{8}$ inch traverse movement to adjust the legs from one notch to the next.

Legs are oval in shape, giving them a very neat appearance and great strength for the weight of metal used.

Mechanics will find this line of calipers and dividers have the QUICKEST ADJUSTMENT, the MOST SENSITIVE ADJUSTMENT, the BEST BALANCE and FINEST FEELING in the hand and on the work, of any line of calipers made, as well as a valuable transfer feature illustrated on page 19.

ARE WHAT COUNT IN EVERYTHING

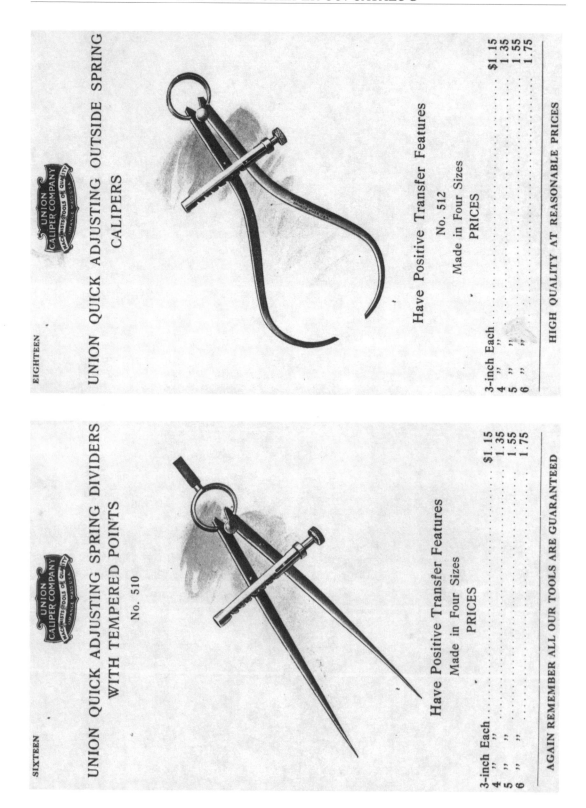

UNION NAIL SETS, No. 30
Every One Warranted

The universal satisfaction our Nail Sets and Punches are giving to users, is due to the fact that right material is used, properly treated by modern methods.

Sizes: A, 1/4 inch body, .060 inch point
B, 1/4 ", .085 "
C, 5/16 ", .105 "
D, 3/8 ", .130 "

PRICES

Per Dozen............$1.00   Per Gross............$12.00

Packed one Dozen in a box.

TRY OUR NAIL SETS AND PUNCHES

ILLUSTRATING POSITIVE TRANSFER FEATURE ON UNION QUICK ADJUSTING SPRING CALIPERS AND DIVIDERS

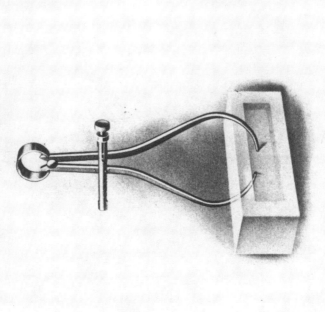

This positive transfer feature mechanics will find very valuable for the measurement over flanges, thickness of panels, internal chambers having flanges which must be measured over or in fact, any place where accurate transfer of sizes must be taken over flanges or shoulders.

To operate, just take the size as you would with any ordinary caliper, getting it exact, note the notch in which the adjusting bar is set, raise the adjusting bar, remove the caliper from the work and set it back to the notch noted. A very simple, positive process. No danger of mistakes.

ALL OUR TOOLS ARE DESIGNED TO GIVE THE MAXIMUM SERVICE

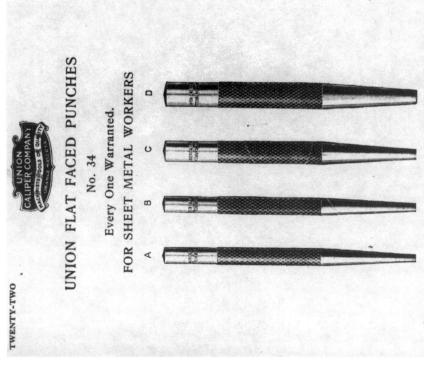

UNION FLAT FACED PUNCHES
No. 34
Every One Warranted.
FOR SHEET METAL WORKERS

These are made of the same high grade steel as our Nail Sets and receive the same treatment; the sizes of bodies and points are also the same.

PRICES
Per Dozen............$1.00 Per Gross............$12.00

Packed One Dozen in a box.

EVERY NAIL SET AND PUNCH WARRANTED

UNION CENTER PUNCHES, No. 32
Every One Warranted

These Punches are made of the same grade of steel as our Nail Sets and receive the same careful treatment.

Sizes: A, 1/4 inch body, .070 inch point
B, 1/4 ", ".093 " "
C, 5/16 ", ".140 " "
D, 3/8 ", ".200 " "

PRICES
Per Dozen............$1.00 Per Gross............$12.00

Packed One Dozen in a box.

YOU WILL SEE HOW MUCH BETTER THEY ARE THAN SOME YOU HAVE USED

1915 UNION CALIPER CO. CATALOG

TWENTY-FOUR

UNION THREAD GAUGE

No. 50

V THREAD

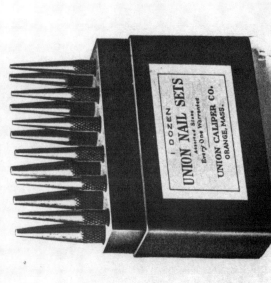

This Gauge has 30 pitches, as follows: 4, 4½, 5, 5½, 6, 7, 8, 9, 10, 11, 11½, 12, 13, 14, 15, 16, 18, 20, 22, 24, 26, 27, 28, 30, 32, 34, 36, 38, 40, 42.

The teeth are accurately cut.

PRICE, No. 50

Price Each..$1.25

ALL THE PITCHES IN COMMON USE FOUND IN THIS GAUGE

TWENTY-THREE

UNION NAIL SETS IN DISPLAY CARTONS

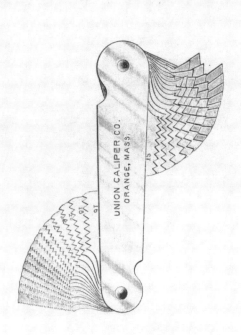

Without extra cost to our dealers, we will ship our Nail Sets, Center Punches, Flat Faced and Prick Punches in these cartons. They are packed assorted dozens in each carton. Remember all our Nail Sets, Center Punches, Flat Faced and Prick Punches are GUARANTEED TO STAND THE ABUSE AND GIVE SATISFACTION.

PRICES

Per Dozen............$1.00 Per Gross............$12.00

FIRST-CLASS GOODS WELL DISPLAYED ARE EASIER SOLD

145

UNION THICKNESS GAUGE
No. 300

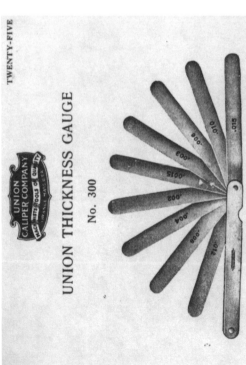

Made in two sizes with 9 Tempered leaves in each size. Each leaf ½ inch in width, thickness as follows: .0015; .002; .003; .004; .006; .008; .010; .012; .015.

Leaves are very flexible and accurate to thickness. The one-half thousandth measurement can be taken on both gauges.

PRICES Each, No. 300 and No. 300 A

No. 300
Case 4⅞ inches long; 9 leaves, each 4½ inches long by ½ inch in width.
Price...$1.50

No. 300 A
Case 3 7/16 inches long; 9 leaves, each 3⅛ inches long by ½ inch in width.
Price...$1.00

THE ONE-HALF THOUSANDTHS CAN BE MEASURED WITH OUR THICKNESS GAUGES

UNION TEMPERED RULES, No. 400

Made in the following Graduations:

No. 1, 1st Cor., 10, 20, 50, 100; 2d, 12, 24, 48; 3d, 16, 32, 64; 4th, 14, 28
No. 4, 1st Cor., 64; 2d, 32; 3d, 16; 4th, 8
No. 7, 1st Cor., 64; 2d, 32; 3d, 16; 4th, 100

Approximate THICKNESS, No. 18 GAUGE

Widths: Inches	½	9/16	⅝	¾	⅞	1	1⅛	1¼	1¼	1¼
Lengths: "	1	2	3	4	6	9	12	18	24	36	48
Prices:	$0.15	.25	.35	.45	.65	1.00	1.25	2.00	2.50	5.00	7.00

UNION TEMPERED RULES, END GRADUATED, No. 410

Made in No. 4 Graduations, same Widths and Thickness as No. 400.

Lengths: Inches	1	2	3	4	6	9	12	18	24
Prices:	$0.15	.25	.35	.45	.65	1.00	1.25	2.00	2.50

CLEAN-CUT ACCURATE GRADUATIONS

1915 UNION CALIPER CO. CATALOG

TWENTY-EIGHT

UNION KEY SEAT RULE BLOCKS
No. 310

These Blocks are made to convert a Steel Rule, Straight Edge or Square Blade into a Key Seat Rule, with which accurate key ways may be laid out. The face of the blocks being radial in shape, accurate lines can be drawn on large shafts or cylinders, carrying their range of usefulness far beyond the blocks that form a right angle with the Rule or Straight Edge. The radial faced shape in no way interferes with their usefulness on the smaller diameters of work.

These Blocks may be also used as Gauges on Rules and Square Blades. They also make in connection with a rule an excellent corner square which is sometimes a handy tool.

These Blocks are hardened. Made in two sizes.

No. 310

For small and medium work, 1 3-16 inches long.
Price, Per Pair .... $.50

No. 310 A

For larger work, 1 1-2 inches long.
Price, Per Pair .... $.75

ALL OUR TOOLS ARE DESIGNED TO GIVE BETTER SERVICE

TWENTY-SEVEN

UNION TEMPERED RULES, FLEXIBLE
No. 420

Graduated on One Side Only

Made of a Very Thin Spring Tempered Stock, about 1/2 inch in width.
Furnished in No. 10 Graduations, 32ds and 64ths.
Furnished in No. 11 Graduations, 64ths and 100ths.

Lengths: Inches   4   6   9   12
Prices: $0.45 .65 1.00 1.25

Leather Cases Furnished with 4-inch and 6-inch.

UNION TEMPERED NARROW RULES
No. 430

About 3/16 inch Wide, No. 18 GAUGE

Spring Tempered, Graduated on One Corner of Each Side.
Furnished in No. 10 Graduations, 32ds and 64ths.
Furnished in No. 11 Graduations, 64ths and 100ths.

Lengths: Inches   4   6   9   12
Prices: $0.45 .65 1.00 1.25

THAT ARE EASY TO READ

1915 UNION CALIPER CO. CATALOG

UNION COMBINATION SQUARES WITH HARDENED BLADES

TWENTY-NINE

A very handy tool. Space does not permit enumerating the many different kinds of work that can be successfully done with our Combination Squares.

Made in Five Sizes

Blades furnished in

	No. 1 Grad.		No. 4 Grad.		No. 7 Grad.	
1st Corner	10, 20, 50, 100	1st Corner	64	1st Corner	64
2d "	12, 24, 48	2d "	32	2d "	32
3d "	16, 32, 64	3d "	16	3d "	16
4th "	14, 28	4th "	8	4th "	100

PRICES, Each

		with Center Head	without
6-inch,		$1.50	$1.00
9 "		1.75	1.25
12 "		2.00	1.50
18 "		2.75	2.25
24 "		3.25	2.75

THAN YOU ARE GETTING IF YOU ARE NOT USING UNION TOOLS

UNION ADJUSTABLE HACK SAW FRAME

No. 500

THIRTY

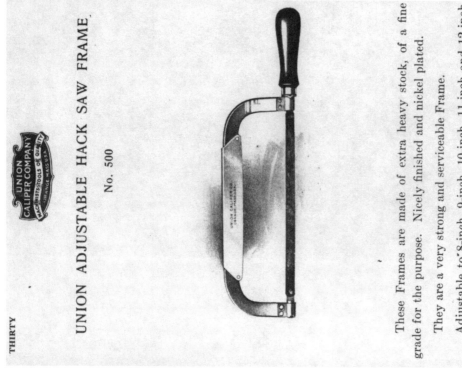

These Frames are made of extra heavy stock, of a fine grade for the purpose. Nicely finished and nickel plated.

They are a very strong and serviceable Frame.

Adjustable to 8-inch, 9-inch, 10-inch, 11-inch and 12-inch saws.

Prices, Each $1.00

IT IS NOT WHAT WE HAVE DONE OR ARE GOING TO DO

THIRTY-ONE

UNION SOLID HACK SAW FRAME
No. 501

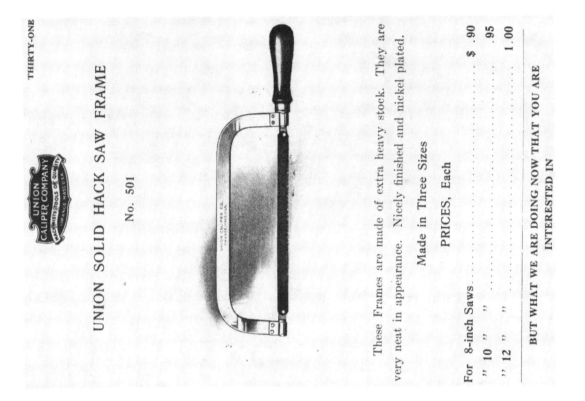

These Frames are made of extra heavy stock. They are very neat in appearance. Nicely finished and nickel plated.

Made in Three Sizes

PRICES, Each

For 8-inch Saws	$ .90
" 10 " "	.95
" 12 " "	1.00

BUT WHAT WE ARE DOING NOW THAT YOU ARE INTERESTED IN

ADDITIONAL MACHINIST'S TOOL PATENTS

Patents listed in this section are additions to the patents included in *American Machinist's Tools, An Illustrated Directory of Patents,* published in 1993. There is at least one drawing for every patent and all are indexed in three separate ways: chronologically by date and patent number; alphabetically by patentee name; and by type of tool.

Most of the added patents issued before 1905 came to light when the tools covered by the patents came to light. Many of the later patents are for test indicators, which were not fully appreciated when the first book was prepared.

Thanks to those of you who wrote or called to point out patents which should be included.

INDEX BY DATE AND PATENT NUMBER

1860

July 17
    Barnett, Samuel   Combined T Square, Protractor and Rule   29133
                                   Used by Darling, Brown & Sharpe

1867

July 2
    Hurd, Ivory A.   Turning Lathe Gage   66239

October 1
    Elmer, D.F.   Index Gage and Caliper   69418

1868

January 7
    Darling, Samuel   Straight-Edge   73082

June 16
    Lane, C.M.   Combination Tool   78974

August 4
    Detrick, Jacob S.   Counting Register   80612

October 27
    Wilkinson, John D. &
        Boyle, E.O.   Measuring Gage   83577

1869

November 23
    Manchester, J.   Pocket Rule   97099
                           Used by Combination Tool Co.

INDEX BY DATE AND PATENT NUMBER

1883

August 21
   Butt, Bozwell B.   Right Angled Level   283564

1885

March 31
   Getty, Fred & Dickinson, Fred   Adjusting Device for Machinist's Tools   314663

1887

January 4
   Wright, Jacob D.   Rapidly Adjusting Nut for Calipers and Dividers   355430
                                        Used by Wright Machine Co.

November 8
   Long, Charles   Spirit Level   372921
                                        Used by R.J. Sanford

1888

August 7
   Almorth, Gustaf   Protractor   387481
                                        Used by Stark Tool Co.

1889

May 14
   Brewer, Adolph   Gage for Grinding Twist or Flat Drills   403175

July 23
   Seaver, Frederic   Centering Device   407617

December 3
   Melick, William B.   Clinometer   16683

1893

September 26
   Weiss, Louis T.   Speed Measure   505582
                                        Used by Weiss Brothers

1894

January 2
   Bates, George A   Self Registering Try Square   511746

1896

June 30
   Smith, Oberlin   Gage (design patent)   25719
                                        Used by Pratt & Whitney

INDEX BY DATE AND PATENT NUMBER

    Strange, Emerson C.   Mechanic's Tool   563089

1898

October 4
    Goddard, John E.   Gaging Implement   611625

November 8
    Benes, Francis   Dividers or Compasses   613814

1901

July 2
    Comstock, Edwin M.   Center Punch and Gage   677339

October 29
    Miller, John C.   Indicator for Lathes   685288

1902

February 18
    Sears, Clarence   Gage   693744

October 9
    Jacobs, Frederick B. &
        Waterhouse, Harold   Test Indicator   715582

1906

January 26
    Peterson, Gustave   Indicator for Lathes   811244

May 8
    Hansen, G.L.   Measurement Indicator   820303

July 17
    Cann, James D.   Combination Gage   826311
        Used by Herbert Dyke

July 31
    Hendrikson, Adolph   Micrometer Attachment for Linear Scales   827443

December 4
    Ennis Benjamin F.   Indicator for Trueing Up Work   837677

1907

February 5
    Sigrist, J.   Surface Indicator   843043

October 29
    Dennis, Thomas L. &
        Lindolm, Arthur   Work Locating Indicator   869483
        Used by Lindholm & Davis

INDEX BY DATE AND PATENT NUMBER

1908

May 19
    Dissell, William   Combined Gage and Calipers   888070

May 26
    Hendrikson, Adolph   Calipers and Dividers   888498

October 6
    Boettcher, Henry   Universal Indicator   900472

December 8
    Roth, G.M.   Gage   906164

1909

April 6
    James, William A.   Surface Indicator   917444
        Used by Sandwich Electric Co.

October 26
    Wheeler, Percey G.   Test Indicator   937978

1910

February 8
    Perkins, William   Marking Gage   948923
        Used by Florence Iron Works Co.

March 29
    Nash, Lewis H.   Measuring Instrument   953282
        Used by Nash Engineering Co.

June 28
    Pember, John B.   Micrometer Attachment for Machinist's Calipers   962583

September 20
    Harris, William H.   Calipers   970817

1911

April 18
    Mack, George   Twist Drill Gage   989657

May 23
    Linnahan, Peter J.   Micrometer Indicator   993084

August 22
    Bowers, George W.   Test Indicator   1001121

October 24
    Erb, Edmund M.   Test Indicator   1006924

November 21
    Walker, Fred   Depth Gage   1009605

1912

March 5
    Zuegner, Lewis W. &
      Langsdon, Al — Universal Test Indicator — 1019400

October 1
    Leigh, Lewis L. — Surface Indicator — 1040210

1913

March 18
    Laplant, Frank — Dial Test Indicator — 1056186

November 18
    Fuchs, Leon — Surface Indicator — 1079169
    Used by Leon Fuchs Mfg. Co. And Bratschi Mfg. Co.

1914

May 26
    Gheen, John W. — Surface Indicator — 1097797

1915

February 23
    Grant, John H. — Indicator for use with Calipers — 1129289

March 23
    Coes, Zorester B. — Micrometer — 1132704

April 6
    Wolfe, Joseph L. — Test Indicator — 1134713
    Used by J.L. Wolfe & Son

May 18
    Warner, Joseph N. — Universal Test Indicator — 1139936

October 26
    Oslund, John E. — Universal Test Indicator — 157800
    Used by Oslund Tool & Die

1916

January 11
    Newmann, H.K. &
      Andrews, H.K. — Test Indicator — 1167592

May 23
    Wheeler, Percey G. — Indicator — 1184399

June 13
    Kinney, Frank B. — Tool Indicator — 1186911

INDEX BY PATENTEE

Almorth, Gustaf
    August 7, 1888   Protractor   387481
        Used by Stark Tool Co.

Barnett, Samuel
    July 17, 1860   Combined T Square, Protractor and Rule   29133
        Used by Darling, Brown & Sharpe

Bates, George A.
    January 2, 1894   Self Registering Try Square   511746

Benes, Francis
    November 8, 1898   Dividers or Compasses   613814

Boettcher, Henry P.
    October 6, 1908   Universal Indicator   900472

Bowers, George W.
    August 22, 1911   Test Indicator   1001121

Brewer, Adolph F.
    May 14, 1889   Gage for Grinding Twist or Flat Drills   403175

Butt, Bozwell B.
    August 21, 1883   Right Angled Level   283564

Cann, James D.
    July 17, 1906   Combination Gage   826311
        Used by Herbert Dyke

Coes, Zorester B.
    March 23, 1915   Micrometer   1132704

Comstock, Edwin M.
    July 2, 1901   Center Punch and Gage   677339

Darling, Samuel
    January 7, 1868   Straight-Edge   73082

Dennis, Thomas L. & Lindholm, Arthur C.
    October 29, 1907   Work Locating Indicator   869483
        Used by Lindholm & Davis

Detrick, Jacob S.
    August 4, 1868   Counting Register   80612

Dissell, William
    May 19, 1908   Combined Gage and Calipers   888070

Elmer, D.F.
    October 1, 1867   Index Gage and Caliper   69418

Ennis, Benjamin F.
    December 4, 1906   Indicator for Trueing Up Work   837677

Erb, Edmund M.
    October 24, 1911   Test Indictor   1006924
        Used by E.M. Erb Mfg. Co.

INDEX BY PATENTEE

Fuchs, Leon
 November 18, 1913 Surface Indicator 1079169
  Used by Leon Fuchs Mfg. Co. and Bratschi Mfg. Co.

Getty, Fred I. & Dickinson, Fred
 March 31, 1885 Adjusting Device for Machinist's Tools 314663

Gheen, John W.
 May 26, 1914 Surface Indicator 1097797

Goodard, John E.
 October 4, 1898 Gaging Implement 611625
  Used by Montgomery & Co.

Grant, John H.
 February 23, 1915 Indicator for use with Calipers 1129289

Hansen, G.L.
 May 8, 1906 Measurement Indicator 820303

Harris, William H.
 September 20, 1910 Calipers 970817

Hendrikson, Adolph F.
 July 31, 1906 Micrometer Attachment for Linear Scales 827443
 May 26, 1908 Calipers and Dividers 888498

Hurd, Ivory A.
 July 2, 1867 Turning Lathe Gauge 66239

Jacobs, Frederic B. & Waterhouse, Harold
 October 9, 1902 Test Indicator 715582

James, William A. Surface Indicator 917444
 April 6, 1909 Used by Sandwich Electric Co.

Kinney, Frank B.
 June 13, 1916 Tool Indicator 1186911

Lane, C.M.
 June 16, 1868 Combination Tool 78974

Laplant, Frank
 March 18, 1913 Dial Test Indicator 1056186

Leigh, Lewis L.
 October 1, 1912 Surface Indicator 1040210

Linnahan, Peter J.
 May 23, 1911 Micrometer Indicator 993084

Long, Charles B.
 November 8, 1887 Spirit Level 372921
  Used by R.J. Sanford

Mack, George
 April 18, 1911 Twist Drill Gage 989857

Manchester, J.
 November 23, 1869 Pocket Rule 97099
  Used by Combination Tool Co.

INDEX BY PATENTEE

Melick, William B.		
December 3, 1889	Clinometer	416683
	Used by Melick Clinometer Co.	
Miller, John C.		
October 29, 1901	Indicator for Lathes	685288
Nash, Lewis H.		
March 29, 1910	Measuring Instrument	953282
	Used by Nash Engineering Co.	
Newman, K.A.& Andrews, H.K.		
January 11, 1916	Test Indicator	1167592
Oslund, John E.		
October 26, 1915	Universal Test Indicator	1157800
	Used by Oslund Tool & Die	
Payler, J.W.		
April 6, 1892	Handle for Files or Other Tools	472140
Pember, John		
June 28, 1910	Micrometer Attachment for Machinist's Calipers	962583
Perkins, William E.		
February 8, 1910	Marking Gage	948523
	Used by Florence Iron Works Co.	
Peterson, Gustave E.		
January 26, 1906	Indicator for Lathes	811244
Roth, G.M.		
December 8, 1908	Gage	906164
Sears, Clarence M.		
February 18, 1902	Gage	693744
Seaver, Frederic A.		
July 23, 1889	Centering Device	407617
Sigrist, J.		
February 5, 1907	Surface Indicator	843043
Smith, Oberlin		
June 30, 1896	Gage (Design Patent)	25719
	Used by Pratt & Whitney Co.	
Strange, Emerson C.		
June 30, 1896	Mechanic's Tool	563089
Walker, Fred J.		
November 21, 1911	Depth Gage	1009605
Warner, Joseph H.		
May 18, 1915	Universal Test Indicator	1139936
Weiss, Louis T.		
September 26, 1893	Speed Measure	505582
	Used by Weiss Brothers	

Wheeler, Percey G.
    October 26, 1909　　　　　Test Indicator　　　　　　　　　　　　　　937978
Used by C.E. Robinson Co.
    May 23, 1916　　　　　　　Indicator　　　　　　　　　　　　　　　1184399
Wilkinson, John D.& Boyle, E.O.
    October 27, 1868　　　　　Measuring Gage　　　　　　　　　　　　83577
Wolfe, Joseph L.
    April 6, 1915　　　　　　　Test Indicator　　　　　　　　　　　　　1134713
    　　　　　　　　　　　　　Used by J.L. Wolfe & Son
Wright, Jacob D.
    January 4, 1887　　　　　Rapidly Adjustable Nut For Calipers and Dividers　　355430
    　　　　　　　　　　　　　Used by Wright Machine Co.
Zuegner, Lewis W.& Langsdon, Al
    March 5, 1912　　　　　　Universal Test Indicator　　　　　　　　　1019400

INDEX BY TOOL TYPE

JOINTED OR SPRING CALIPERS AND DIVIDERS

1868

June 16
    Lane, C.M.　　　　　　　Combination Tool　　　　　　　　　　　　78974
October 27
    Wilkinson, John D. &
    Boyle, E.O.　　　　　　　Measuring Gage　　　　　　　　　　　　　83577

1887

January 4
    Wright, Jacob D.　　　　　Rapidly Adjusting Nut for Calipers and Dividers　　355430
    　　　　　　　　　　　　　Used by Wright Machine Co.

1898

November 8
    Benes, Francis　　　　　　Dividers or Compasses　　　　　　　　　　613814

1908

May 26
    Hendrikson, Adolph　　　　Calipers and Dividers　　　　　　　　　　　888498

INDEX BY TOOL TYPE

1910

June 28
   Pember, John B.   Micrometer Attachment for Machinist's Calipers   962583

September 20
   Harris, William H.   Calipers   970817

BEAM CALIPERS, TRAMMELS AND BEAM COMPASSES

1898

October 4
   Goddard, John E.   Gaging Implement
611625

1901

July 2
   Comstock, Edwin M.   Center Punch and Gage   677339

CENTERING TOOLS AND GAGES

1889

July 23
   Seaver, Frederic   Centering Device   407617

INDICATORS AND THICKNESS GAGES

1901

October 29
   Miller, John C.   Indicator for Lathes   685288

1902

February 18
   Sears, Clarence M.   Gage
693744

October 9
   Jacobs, Frederic B. &
Waterhouse, Harold   Test Indicator   715582

1906

January 26
    Peterson, Gustave   Indicator for Lathes   811244

May 8
    Hansen, G.L.   Measurement Indicator   820303

December 4
    Ennis, Benjamin F.   Indicator for Trueing Up Work   837677

1907

February 5
    Sigrist, J.   Surface Indicator   843043

October 29
    Dennis, Thomas L. &
    Lindholm, Arthur   Work Locating Indicator   869483
        Used by Lindholm & Davis

1908

October 6
    Boettcher, Henry   Universal Indicator   900472

1909

April 6
    James, William A.   Surface Indicator   917444
        Used by Sandwich Electric Co.

October 26
    Wheeler, Percey G.   Test Indicator   937978
        Used by C.E. Robinson Co.

1911

May 23
    Linnahan, Peter J.   Micrometer Indicator   993084

August 22
    Bowers, George W.   Test Indicator   1001121

October 24
    Erb, Edmund M.   Test Indicator   1006924
        Used by E.M. Erb Mfg. Co.

1912

March 5
    Zuegner, Lewis W. &
    Langsdon, Al   Universal Test Indicator   1019400

INDEX BY TOOL TYPE

October 1
   Leigh, Lewis L.   Surface Indicator   1040210

1913

March 18
   Laplant, Frank   Dial Test Indicator   1056186

November 18
   Fuchs, Leon   Surface Indicator   1079169
                                       Used by Leon Fuchs Mfg. Co and Bratschi Mfg. Co.

1914

May 26
   Gheen, John W.   Surface Indicator   1097797

1915

February 23
   Grant, John H.   Indicator for use with Calipers   1129289

April 6
   Wolfe, Joseph L.   Test Indicator   1134713
                                       Used by J.L. Wolfe & Son

May 18
   Warner, Joseph N.   Universal Test Indicator   1139936

October 26
   Oslund, John E.   Universal Test Indicator   1157800
                                   Used by Oslund Tool & Die

1916

January 11
   Newmann, H.K. &
     Andrews, H.K.   Test Indicator   1167592

May 23
   Wheeler, Percey G.   Indicator   1184399

June 13
   Kinney, Frank B.   Tool Indicator   1186911

LEVELS

1883

August 21
   Butt, Bozwell B.   Right Angled Level   283564

INDEX BY TOOL TYPE

1887

November 8
  Long, Charles B.   Spirit Level   372921
                      Used by R.J. Sanford

1889

December 3
  Melick, William B.   Clinometer   416683
                      Used by Melick Clinometer Co.

MICROMETER CALIPERS AND MEASURING MACHINES

1867

October 1
  Elmer, D.F.   Index Gage and Caliper   69418

1910

March 29
  Nash, Lewis H.   Measuring Instrument   953282
                      Used by Nash Engineering Co.

1915

March 23
  Coes, Zorester B.   Micrometer   1132704

PROTRACTORS AND CIRCLE DIVIDERS

1860

July 17
  Barnett, Samuel   Combined T Square, Protractor, and Rule   29133
                      Used by Darling, Brown & Sharpe

1888

August 7
  Almorth, Gustav   Protractor   387481
                      Used by Stark Tool Co.

INDEX BY TOOL TYPE

1906

July 17
   Cann, James D.
       Combination Gage   826311
       Used by Herbert Dyke

1910

February 8
   Perkins, William
       Marking Gage   948523
       Used by Florence Iron Works Co.

RULES

1868

January 7
   Darling, Samuel
       Straight-Edge   73082

1869

November 23
   Manchester, J.
       Pocket Rule   97099
       Used by Combination Tool Co.

1906

July 31
   Hendrikson, Adolph
       Micrometer Attachment for Linear Scales   827443

SPEED INDICATORS

1868

August 4
   Detrick, Jacob S.
       Counting Register   80612

1893

September 26
   Weiss, Louis T.
       Speed Measure   505582
       Used by Weiss Brothers

INDEX BY TOOL TYPE

SQUARES AND BEVELS

1894

January 2
  Bates, George A.   Self Registering Try Square   511746

1908

May 19
  Dissell, William   Combined Gage and Calipers   888070

SURFACE GAGES

1908

December 8
  Roth, G.M.   Gage   906164

DRILL GAGES

1889

May 14
  Brewer, Adolph   Gage for Grinding Twist or Flat Drills   403175

1896

June 30
  Strange, Emerson   Mechanic's Tool   563089

1911

April 18
  Mack, George   Twist Drill Gage   989657

DEPTH GAGES AND HEIGHT GAGES

1867

July 2
  Hurd, Ivory A.   Turning Lathe Gage   66239

1885

March 31
    Getty, Fred & Dickinson, Fred   Adjusting Device for Machinist's Tools   314663

1911

November 21
    Walker, Fred   Depth Gage   1009605

On the following pages are the patent drawings that were listed in this chapter. The drawings are arranged in chronological order.

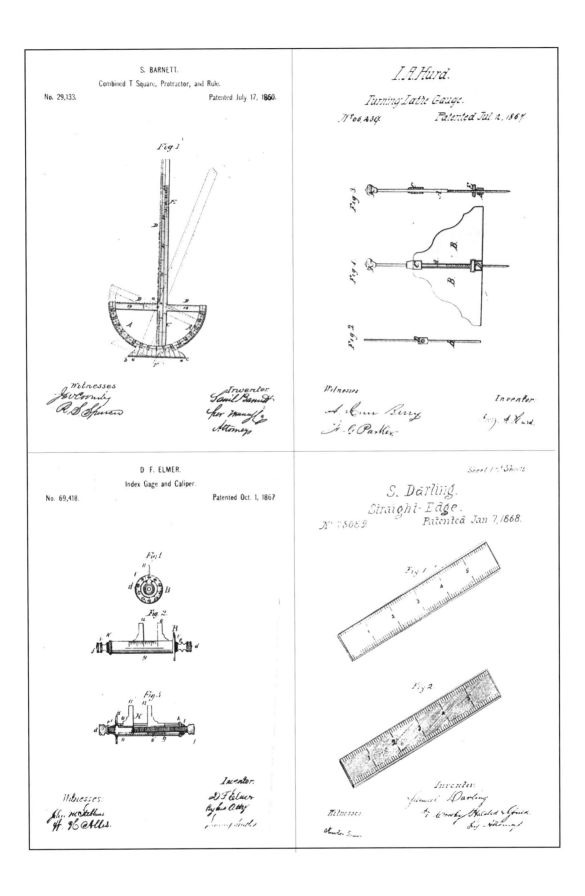

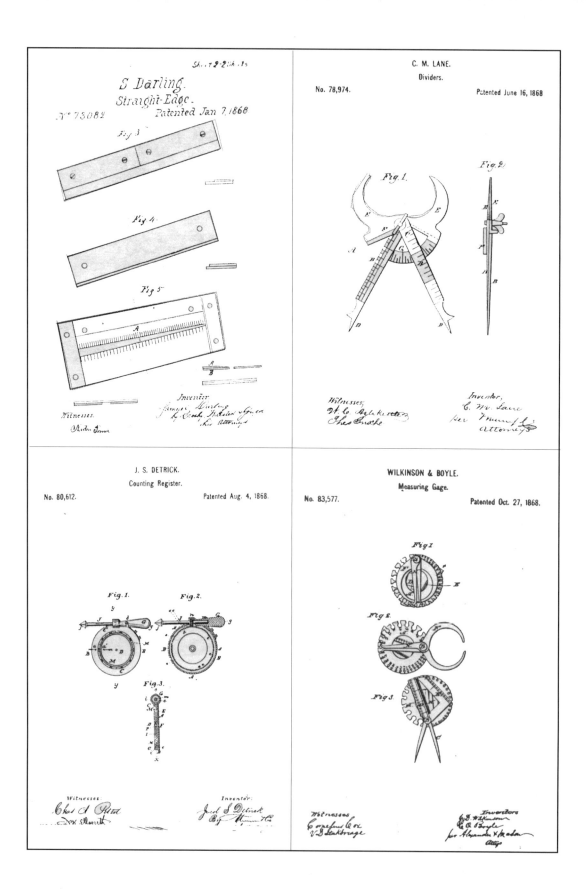

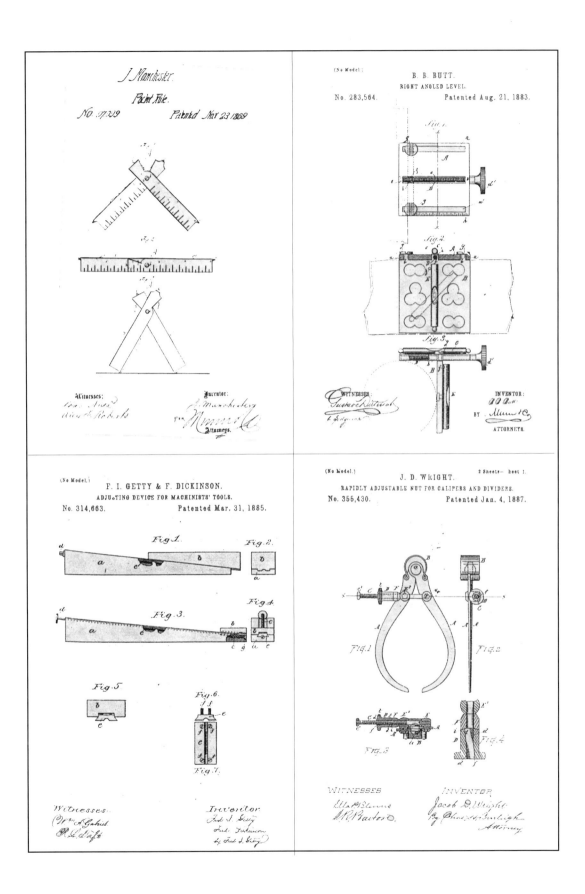

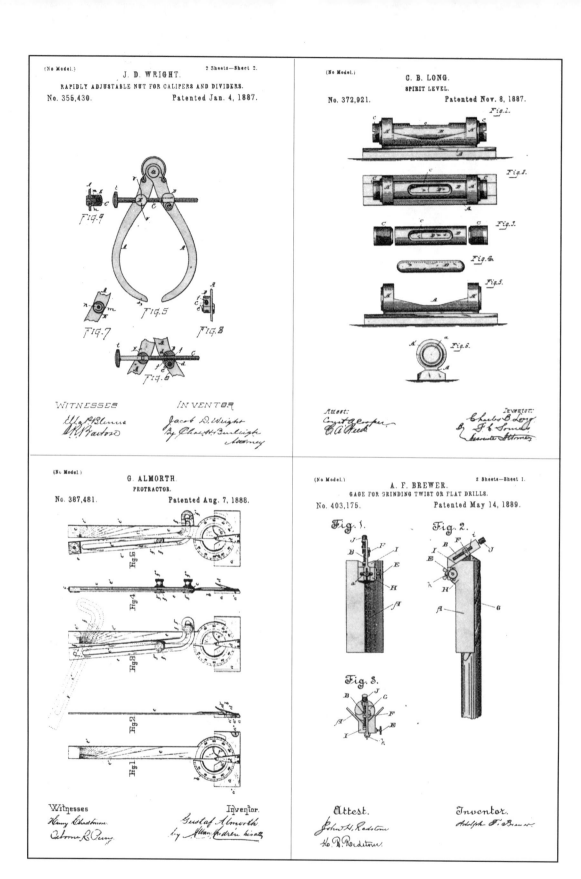

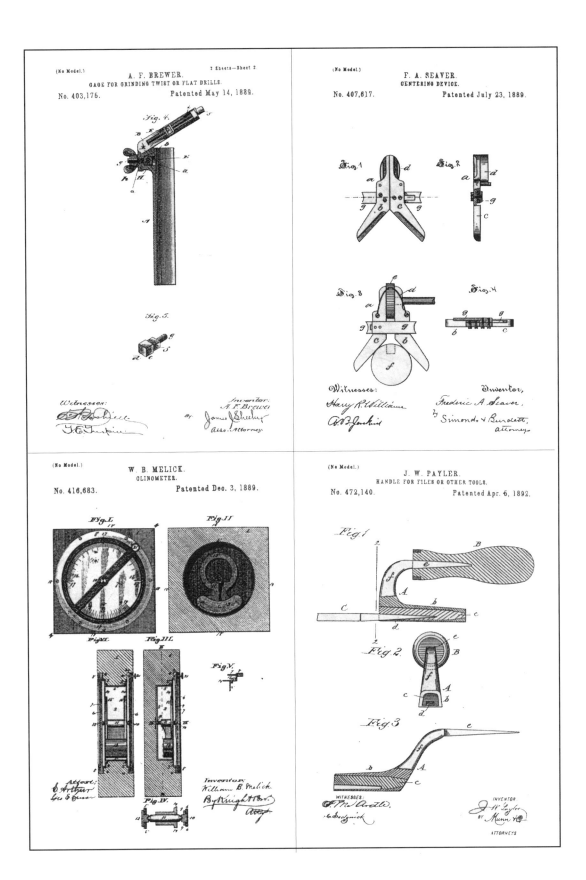

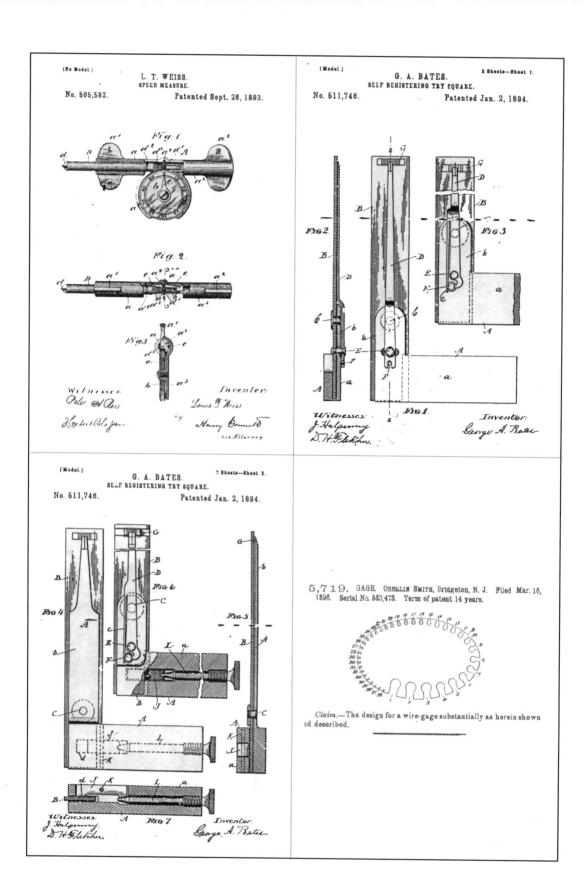

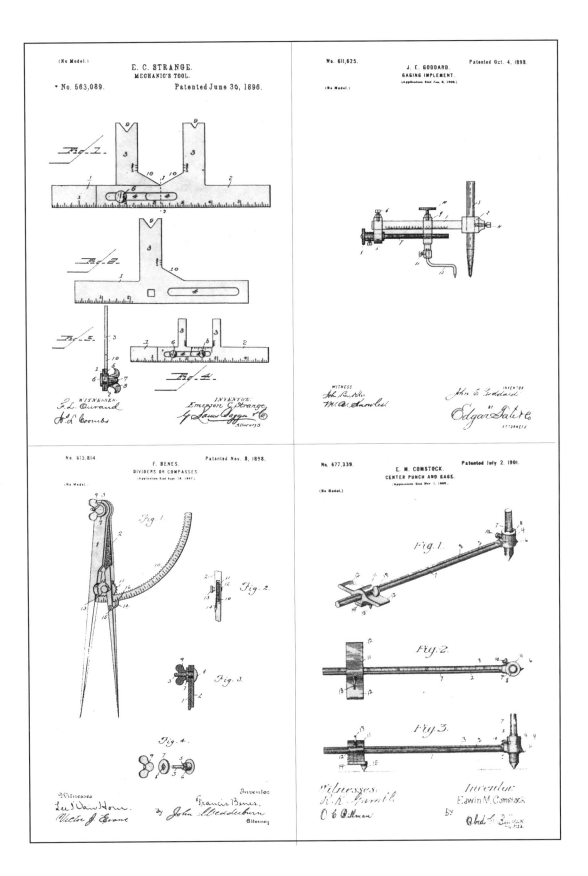

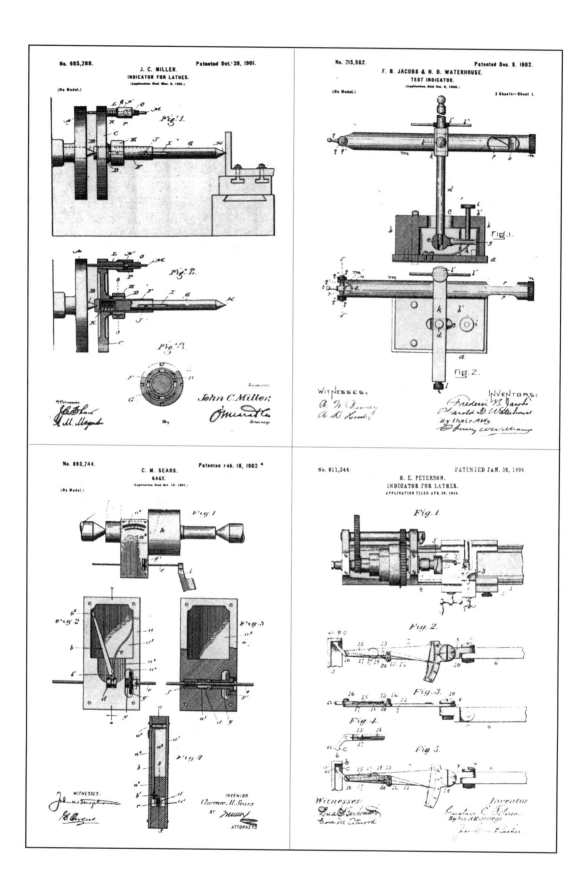

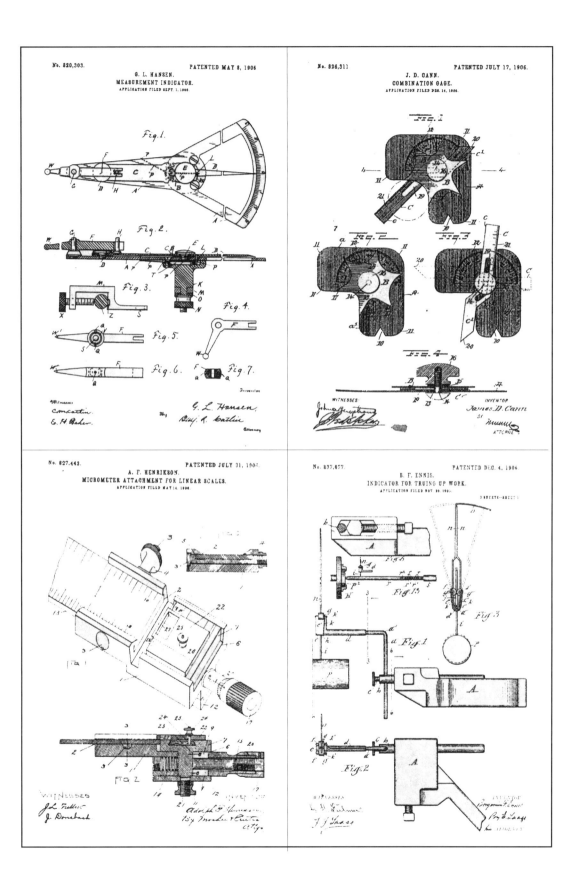

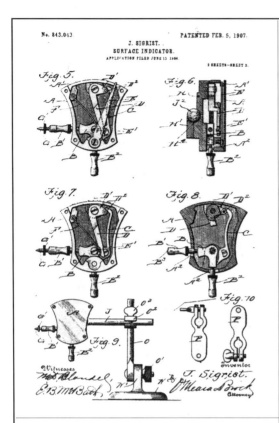
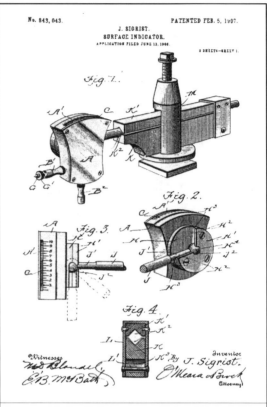
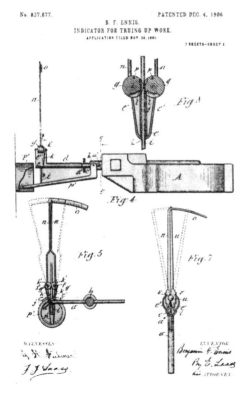
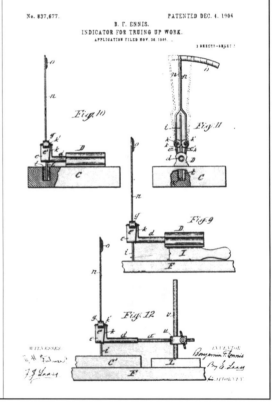

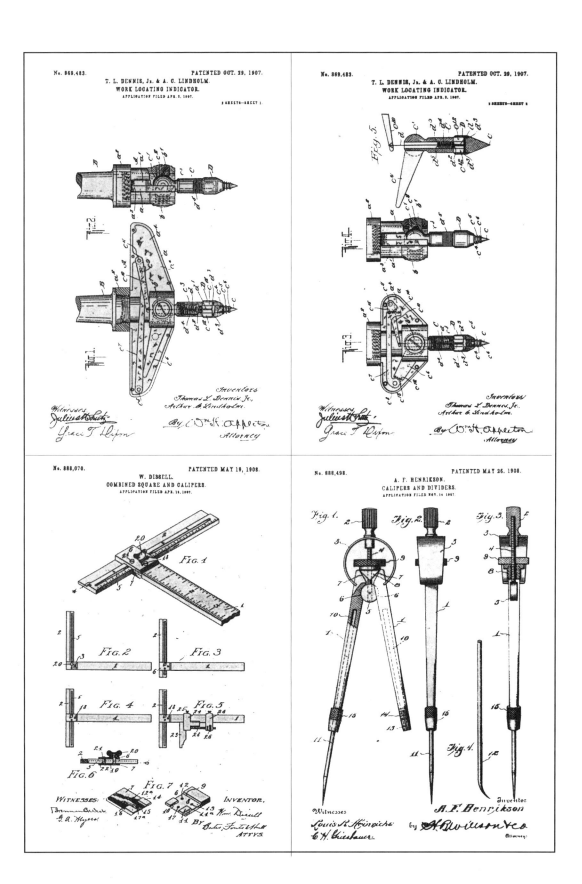

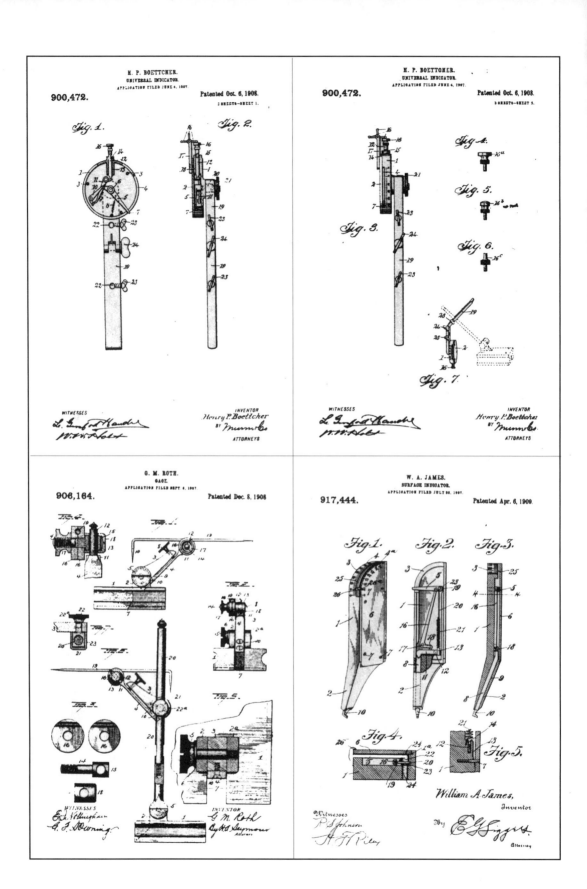

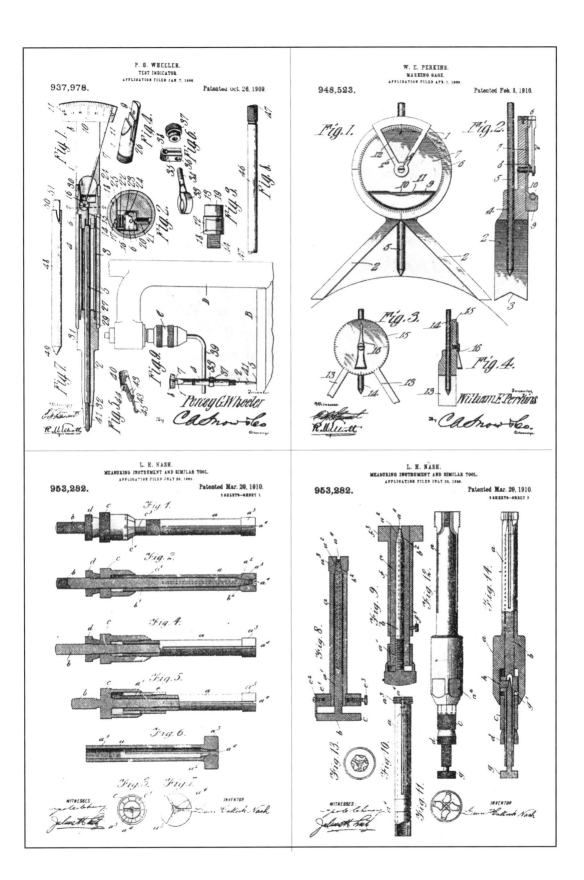

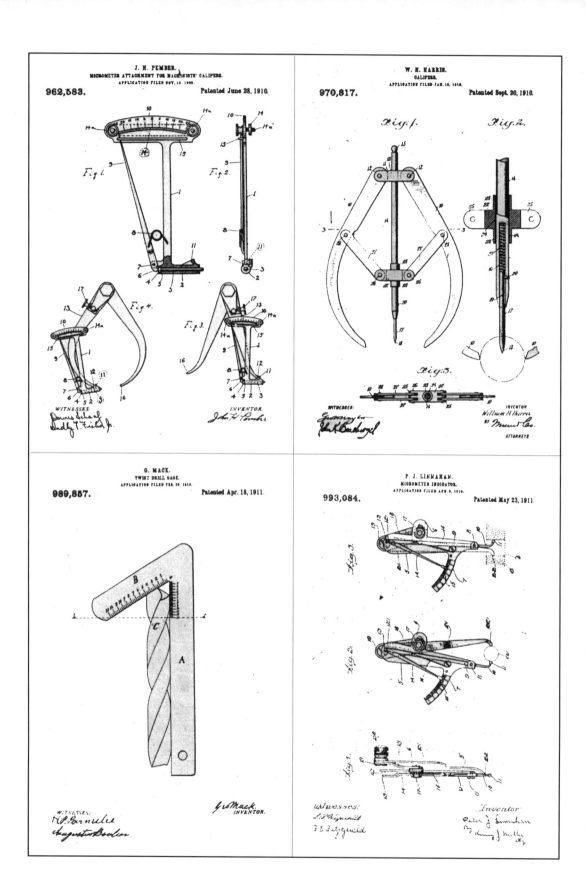

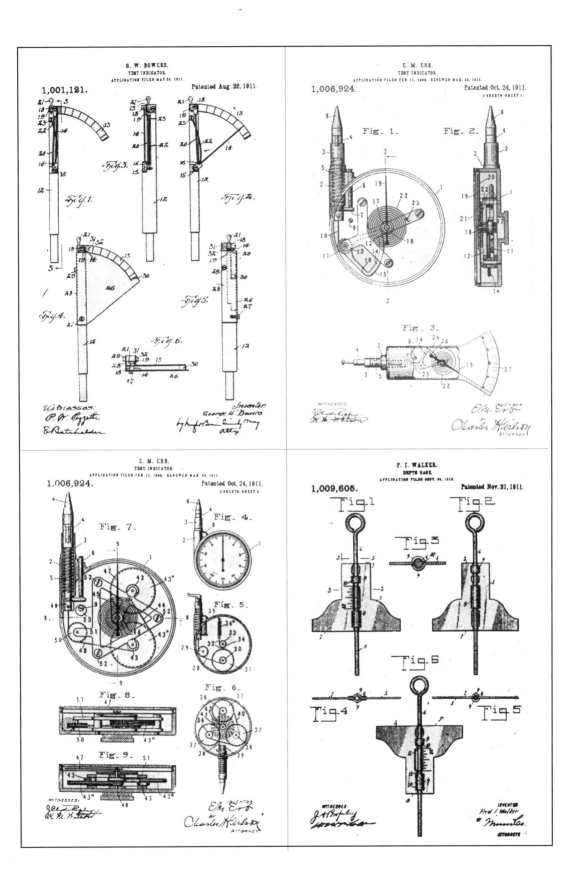

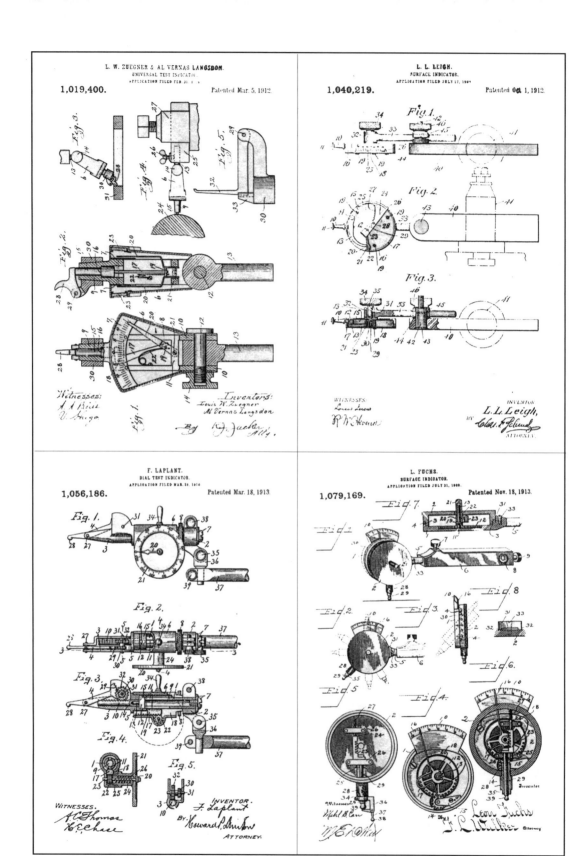

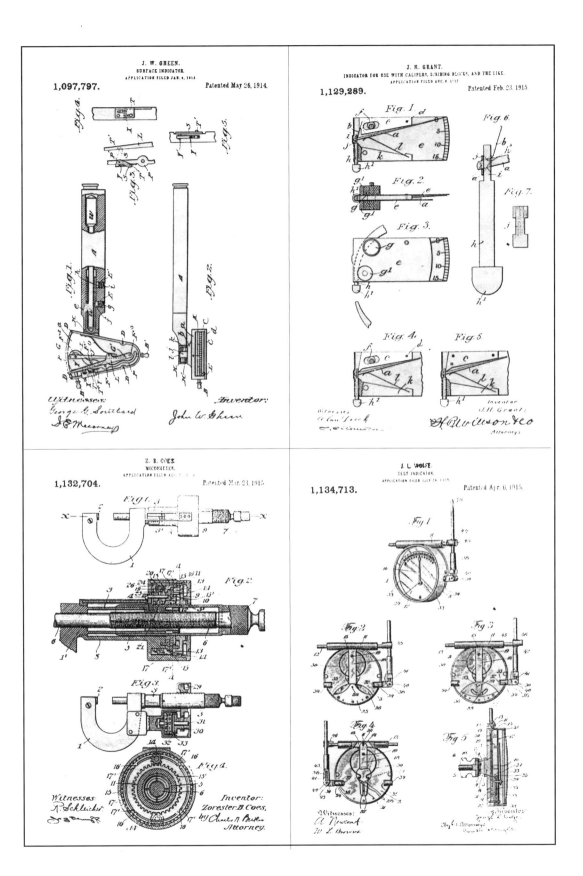

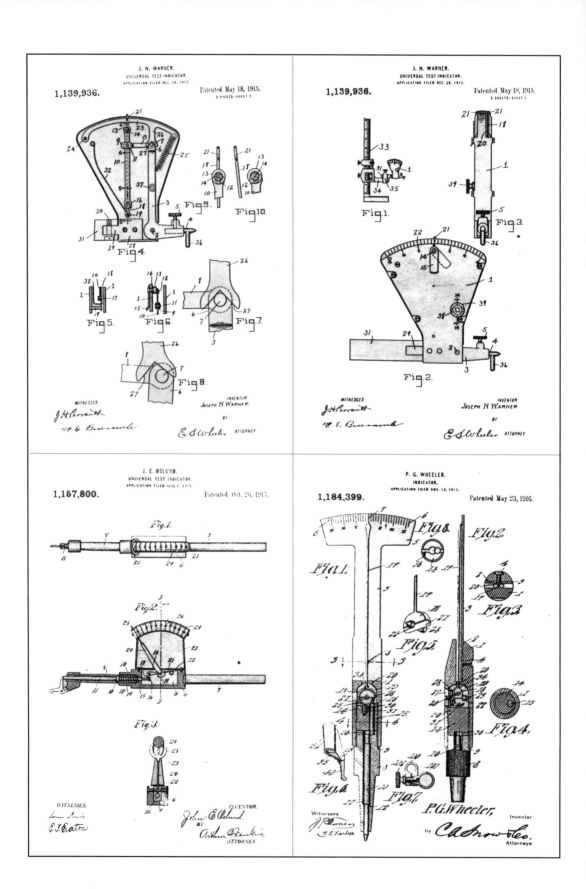

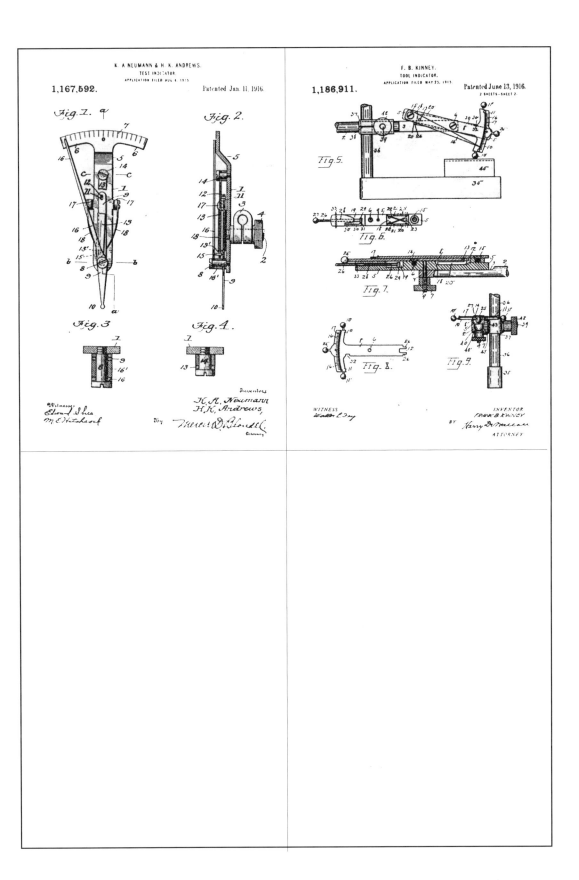